ARCHITECTURE AND ADAPTATION

Architecture and Adaptation discusses architectural projects that use computational technology to adapt to changing conditions and human needs. Topics include kinetic and transformable structures, digitally driven building parts, interactive installations, intelligent environments, early precedents and their historical context, socio-cultural aspects of adaptive architecture, the history and theory of artificial life, the theory of human–computer interaction, tangible computing, and the social studies of technology. Author Socrates Yiannoudes proposes tools and frameworks for researchers to evaluate examples and tendencies in adaptive architecture. Illustrated with more than 50 black and white images.

Socrates Yiannoudes is a Lecturer at the School of Architectural Engineering, Technical University of Crete, Greece. He teaches Architectural Design and Digital Technologies, theory and practice, at undergraduate and postgraduate level. His research interests include digital culture and architecture as well as the application of science and technology studies in architectural research.

ARCHITECTURE AND ADAPTATION

From Cybernetics to
Tangible Computing

Socrates Yiannoudes

Routledge
Taylor & Francis Group
NEW YORK AND LONDON

First published 2016
by Routledge
711 Third Avenue, New York, NY 10017

and by Routledge
2 Park Square, Milton Park, Abingdon, Oxon OX14 4RN

Routledge is an imprint of the Taylor & Francis Group, an informa business

© 2016 Taylor & Francis

The right of Socrates Yiannoudes to be identified as author of this work has been asserted by him in accordance with sections 77 and 78 of the Copyright, Designs and Patents Act 1988.

All rights reserved. No part of this book may be reprinted or reproduced or utilised in any form or by any electronic, mechanical, or other means, now known or hereafter invented, including photocopying and recording, or in any information storage or retrieval system, without permission in writing from the publishers.

Trademark notice: Product or corporate names may be trademarks or registered trademarks, and are used only for identification and explanation without intent to infringe.

Library of Congress Cataloguing in Publication Data
Yiannoudes, Socrates, author.
Architecture and adaptation : from cybernetics to tangible computing / Socrates Yiannoudes.
    pages cm
  Includes bibliographical references and index.
  1. Architecture – Human factors. 2. Architecture and technology.
  3. Smart materials in architecture. 4. Architecture – Technological innovations. I. Title.
  NA2542.4.Y39 2016
  720´.47–dc23   2015027889

ISBN: 978-1-138-84315-8 (hbk)
ISBN: 978-1-138-84316-5 (pbk)
ISBN: 978-1-315-73111-7 (ebk)

Acquisition Editor: Wendy Fuller
Editorial Assistant: Grace Harrison
Production Editor: Ed Gibbons

Typeset in Bembo
by HWA Text and Data Management, London

CONTENTS

List of Figures   *vi*
Foreword   *ix*
Preface   *xii*

Introduction   1

1   Architecture, Adaptation and the Cybernetic Paradigm   9

2   Intelligence in Architectural Environments   55

3   Psycho-Social Aspects   92

4   Contextualizing Interactive Architecture   121

5   Tangible Computing: Towards Open-Potential Environments   172

Bibliography   *205*
Index   *225*

FIGURES

1.1	First-order cybernetic system	12
1.2	Homeostatic system in buildings	12
1.3	Second-order cybernetic system	14
1.4	Le Corbusier "Maison Loucheur, 1929, Plan FLC 18238"	15
1.5	*House 1990*, View of Installation, Archigram	20
1.6	*Suitaloon & Cushicle* in Reclining Position, Michael Webb	21
1.7	Control and Choice, *Timeline*, Peter Cook	22
1.8	The Environment-Bubble, François Dallegret	24
1.9	Anatomy of a Dwelling, François Dallegret	25
1.10	*Fun Palace*: typical plan, Cedric Price	27
1.11	*Fun Palace*: section, Cedric Price	27
1.12	*Ville Spatiale*, Yona Friedman	30
1.13	*Flatwriter*, Yona Friedman	30
1.14	"Organisational Plan as Programme," from the minutes of the Fun Palace cybernetics committee meeting	32
1.15	The structure provides sufficient transparency, from "Architecture in Greece" 3 (1969), Takis Zenetos	35
1.16	Suspended City. The environment. The urban structure co-exists with nature, from "Architecture in Greece" 4 (1970), Takis Zenetos.	35
1.17	Vegetation extending over two levels, from "Architecture in Greece" 8 (1974), Takis Zenetos	36
1.18	Immaterial Architecture (distant future), from "Architecture in -Greece" 8 (1974), Takis Zenetos	36
1.19	Tele-processing cell, from "Architecture in Greece" 8 (1974), Takis Zenetos	37

1.20	Twin-cell envelope, of variable transparency, for two tele-activities assemblies, from "Architecture in Greece" 8 (1974), Takis Zenetos.	37
1.21	Kinetic mechanisms	39
1.22	Section of Bengt Sjostrom Starlight Theatre, Studio Gang	40
1.23	Interior image of Bengt Sjostrom Starlight Theatre, Studio Gang	41
1.24	Interior image of Bengt Sjostrom Starlight Theatre, Studio Gang	41
2.1	*Generator*: view of working electronic model between 1976 and 1979, Cedric Price	56
2.2	*Generator*: initial design network showing three starting points, Cedric Price	57
2.3	Diagram of subsumption architecture.	61
2.4	Sematectonic Lattice, Pablo Miranda Carranza	64
2.5	Equidistribution, Swarm-Roof	65
2.6	Self-assembly of one group, Swarm-Roof	65
2.7	Self-assembly of two groups, Swarm-Roof	66
2.8	*Flight Assembled Architecture*, Gramazio & Kohler and Raffaello D'Andrea	67
2.9	*Flight Assembled Architecture*, Gramazio & Kohler and Raffaello D'Andrea	67
2.10	*Flight Assembled Architecture*, Gramazio & Kohler and Raffaello D'Andrea	68
2.11	*Flight Assembled Architecture*, Gramazio & Kohler and Raffaello D'Andrea	68
2.12	Diagram of Intelligent Environment.	70
2.13	Day view looking into production line, Grimshaw, Rolls-Royce Manufacturing Plant and Headquarters	72
2.14	Living room with robot Astrid, Studio Gang	77
2.15	Fridge with the transportation table, Studio Gang	78
2.16	Living room with the transportation table, Studio Gang	78
3.1	Display of moving geometric shapes	96
3.2	*Animaris Percipiere*, Theo Jansen	97
3.3	*Animaris Rhinoceros*, Theo Jansen	98
3.4	*Kismet,* Science & Technology, robot, Cynthia Breazeal	104
3.5	*Hylozoic Ground* detail, PBAI	107
3.6	*Hylozoic Ground*, PBAI	107
3.7	*Hylozoic Ground*, PBAI	108
3.8	*Hylozoic Ground*, PBAI	109
4.1	Diagram of responsive system	123
4.2	*4D-Pixel,* Studio Roosegaarde	123
4.3	*4D-Pixel,* Studio Roosegaarde	124
4.4	*4D-Pixel,* Studio Roosegaarde	124
4.5	*Flow 5.0*, Studio Roosegaarde	125
4.6	*Flow 5.0*, Studio Roosegaarde	125

4.7	Parametric control of the *SpiritGhost 04* study model	127
4.8	Project *SpiritGhost 04* visualization of the sensponsive classroom's interior space during different lecture modes	128
4.9	*Interactive Wall* in action	129
4.10	*Interactive Wall* in action	129
4.11	*Interactive Wall* in action	130
4.12	Ruairi Glynn, *Performative Ecologies*	138
4.13	Ruairi Glynn, *Performative Ecologies*	138
4.14	Ruairi Glynn, *Performative Ecologies*	139
4.15	Stelarc, *Fractal Flesh*	148
4.16	Stelarc, *Handswriting*	149
5.1	*Reconfigurable House*, Usman Haque and Adam Somlai-Fischer	179
5.2	*Reconfigurable House*, Usman Haque and Adam Somlai-Fischer	180
5.3	*Reconfigurable House*, Usman Haque and Adam Somlai-Fischer	180
5.4	*Reconfigurable House*, Usman Haque and Adam Somlai-Fischer	181
5.5	The "digital layer" in the system of layers analysis of buildings	190

FOREWORD

It's a long walk from antique automata to drone-assembled foam blocks and inhabited "muscles," but this is the *promenade architecturale* on which Socrates Yiannoudes invites us. At issue is a broad ontological question: where is the dividing line between the animate and the inanimate, and where should we situate architecture with respect to this line? With one foot in science fiction and the other solidly planted in real-time reality, Yiannoudes takes seriously what have often been seen as fringe concerns – avant-garde or just in left field – in architectural thought.

What makes this book special and perhaps unique is that Yiannoudes is not just an observer, but an actual participant in the movement he describes. At once scholarly and engaged, the book tries to help us understand both the *longue durée* tendencies and the current potential of approaches that explore the relationship between the living world and architecture.

A central image is that of the "conversation," or the notion of "open-ended reciprocal exchange." Conversation, as the cyberneticist Gordon Pask affirms, is a "process of conjoint concept execution, by participants A and B, as a result of which the agreed part of the concept is distributed or shared and cannot be functionally assigned 'to A' or 'to B'." This image highlights the central question of our relationship to machines, and by extension to so-called intelligent environments: is there, or can there be, any sort of common ground between our artificial constructs and ourselves, as agents, actants, participants in a "conversation"?

It is perhaps not accidental that one of the first "conversations" to be re-conceptualized in terms of artificial intelligence was psychoanalytic. A primitive computer program called ELIZA, written by Joseph Weizenbaum at MIT between 1964 and 1966, during the heyday of AI optimism about the potential of natural-language processing, simulated the conversational responses of a psychoanalyst to affirmations or questions typed in by a human "patient." The results were extraordinarily fertile, given the very basic methods and parameters undergirding the program. Many users (including myself) have been intrigued and disturbed by the responses generated by ELIZA. ELIZA is able to simulate convincingly a therapist's approach, because the therapist–patient relationship (in this model of non-directive therapy) requires very little common ground, and the conversational model puts almost all of the burden of information on the "patient." ELIZA's responses are therefore not required, in order to appear human, to draw on a shared context of experience.

Extrapolating from this example, one of the key insights of this book is to underline the importance of asymmetry in the relationship between human and machinic actors: is "conversation" necessarily symmetrical? Yiannoudes affirms that it is not, and that in fact the most productive interactive relations, in general, are more likely to be asymmetrical. This does not mean that AI and A-Life should not be ambitious in the production of adaptive environments; but their ambitions should not be formally attached to symmetry in terms of shared contextual information. The future of our efforts to make environments intelligent and adaptive does not, in fact, depend on our ability to force upon them our own characteristics.

Perhaps the other crucial notion in this book, which subtly threads through the text, is that of "affordance." The interaction between humans and architecture is no doubt best conceptualized in terms of this complex concept. "Affordance" is a quality of an object or environment that permits or invites certain manipulations – for instance, the handle of a teacup is an invitation to hold it, pick it up, and drink from it. Some teacups do not have handles, notably in the Japanese tradition, in which case the affordance of the teacup is less directive, more intuitive. In the same vein, the affordances of the environment can be quite broad and indirect: how to characterize, for example, the affordances of a public square? Its benches, ground surface, borders and trees are so many ambiguous affordances, allowing for but not determining any particular behavior. This is just as true of the machinically- or electronically-enhanced domestic environment, at least in the viewpoint of this book. Interaction with an intelligent environment, as Yiannoudes sees things, is "open-ended," meaning indeterminate and subject to the effects of learning on the part of both user and environment. Yiannoudes is resolutely in favor of pragmatic advances rather than flashy, one-off coups: following Malcom McCullough, he prefers "the affordances of everyday life rather than fashionable statements in form."

The reader will find this book to be a reliable guide to the last five decades of reflection on interactions between architecture and cybernetics, between living users and semi-autonomous environments, between fantasy and reality, in the matter of AI and A-Life. At ease with both the technical and philosophical sides of the question, Yiannoudes is uniquely suited for this work. He is capable of explaining the most arcane configuration of sensors and actuators, while in a subsequent paragraph summarizing the phenomenological tradition in philosophy.

Despite its welcome and thorough interrogation of architectural flights of fancy from Cedric Price to Kas Oosterhuis, this book comes back, again and again, to the potential of interactive environments for real-world advances in the service of users. It sees ambient intelligence as an "extra layer" – and a needed one – for architecture, just as, more than a century ago, HVAC and other mechanical systems began to be adjoined to domestic and public space.

Architecture and Adaptation is an appropriate title for a work that is itself supple and adaptive, with serene descriptions of the most far-fetched conceptual projects, combined with level-headed assertions about what is most needed in terms of architectural values. Yiannoudes looks forward to a new generation of architectural work – not so far from where we are now – when so-called new technologies will be simply incorporated in our everyday architectural life; when it will not be a question of if, but of how, the conversation between us and architecture, our "inanimate" twin, will be consummated.

David Vanderburgh
Professor of Architecture
University of Louvain

PREFACE

I became interested in the relation between architecture and information technology when I moved to London for my postgraduate studies in 1999, and, in particular, when I saw a presentation by performance artist Stelarc, and a lecture by invited architects Kas Oosterhuis and Ole Bauman at University College London. During my five-year stay in London I came across several related happenings, lectures, shows and events that further affected my thinking about architecture, in the context of a rapidly growing digitally mediated environment. Some experimental architects and students were already producing designs and actual structures that introduced intelligent systems in architectural space, including sensor networks, processing units and actuator devices, and they presented their work in conferences, shows and workshops around the world. As my curiosity for this theme increased, I decided to explore it further in my postgraduate studies. In my PhD I suggested conceptual tools for architectural spaces that are mediated by computer technology, and examined both past and present examples. I synthesized theoretical material from several disciplines, from artificial intelligence to the social studies of science and technology, in order to contextualize this emerging architectural genre. The current work builds on this early study and adds more contributions by rethinking several assumptions and critiques, discusses past and recent projects to keep up with the current state of the art, and expands the theoretical material.

Overall, the book deals with buildings, structures and spaces that make use of ubiquitous computing systems. It starts with a historical review of relevant precedents, to suggest possible links between the recent past and current projects, including digitally-driven structures, interactive installations and intelligent environments. Particular emphasis is given on the cybernetic concept of adaptation, a vision that can be traced back to the experimental work of the 1960s avant-garde. Relevant insights from the fields of the social studies of technology, psychology and the history and theory of artificial intelligence are further discussed, in order to address the psychological and social aspects of augmented environments and ambient intelligence. Questions about interactivity are also examined, employing concepts and ideas from the theory of human–computer interaction, cyberculture and post-cognitivist theories. At the same time, the book highlights concerns about the impact on the patterns of domestic life, as well as spatial and functional transformations in domestic space.

The book has an interdisciplinary outlook. Thus, it attempts to construct a coherent conceptual framework that could help researchers and academics assess relevant projects, current tendencies and work towards future design and implementation. Therefore, the book is aimed at students, professionals and academics interested in the link between architecture and ubiquitous computing applications, or more broadly in spaces and artifacts mediated by intelligent systems. It aspires to become a platform for a critical debate for further research and development, bringing together architects, artists and engineers.

Very few books make such an extensive contribution to the subject matter. There are three main reasons for this: First, the book looks at the early history, in order to propose useful links. No other book makes this connection so constructive and critical to the discussion. Second, it has an interdisciplinary nature; it synthesizes ideas, techniques and projects from diverse fields, ranging from architectural theory to computer science and psychology. Third, and most important, it is the only book so far to my knowledge that contextualizes augmented architecture within the social studies of technology, post-cognitivist theories, and the theoretical studies of human–computer interaction, to propose an interdisciplinary framework.

The book would not exist without the contribution of several people who were involved in its realization. I would like to thank the Routledge publication committee for approving my proposal. Of course it would not have been possible without the help of my Routledge editors, Grace Harrison and Wendy Fuller, whom I thank for their professionalism and constant guidance. I would also like to thank David Vanderburgh for taking the time to read the draft, comment, and write the Forward. For granting me permission to publish copyrighted projects I would like to thank: The Archigram Archives, Dennis Crompton and Shelley Power, The Canadian Center for Architecture and Caroline Dagbert, Stelarc, Alessandro Saffioti, Ruairi Glynn, Daan Roosegaarde and Anne Vlaanderen, Jan Willmann, Fabio Gramazio, Matthias Kohler and Raffaello D'Andrea, Theo Jansen, François Dallegret, Pablo Miranda Carranza, Edmund Sumner and Sophia Gibb, Salvador Miranda and Philip Beesley, Rachael Robinson and the MIT Museum, the Hellenic Institute of Architecture, Elias Constantopoulos and Marianna Milioni, Studio Gang Architects, Gr egg Garmisa and Sarah Kramer, Usman Haque and Adam Somlai-Fischer, Yona Friedman and Jean-Baptiste Decavèle, Fondation Le Corbusier and Delphine Studer, Sophia Karabinou and OSDEETE, Caroline Chatelain and ADAGP, and Konstantinos-Alketas Oungrinis. Finally I would like to thank all those people, friends, colleagues, and family who supported my work, and contributed in one way or another to this project.

INTRODUCTION

In his essay "A Home is Not a House" (1965), Reyner Banham proposed the speculative "unhouse," an almost invisible house reduced to its environmental systems and electromechanical domestic services and functions.[1] A decade later, Nicholas Negroponte, in his book *Soft Architecture Machines*, discussing the emergence of what he called "responsive architecture" in the context of the prospects of machine intelligence, suggested that "maybe a house is a home only once it can appreciate your jokes."[2] His point was that it may be possible to create intelligent environments able to make decisions and proactively execute functions, according to their internal computations. Banham's "unhouse" and Negroponte's "responsive architecture" were visions of houses capable of responding to the desires of their occupants by means of their information and energy management devices. These concerns revived in the 1990s, partly because of the wide dissemination of computational technologies. Kas Oosterhuis's[3] *e-motive* architecture, illustrated by his speculative *E-motive House*[4] and *Trans_PORTs*[5] projects, were proposed as sensor–actuator equipped flexible structures, able to exchange information with their local and distant environment. *E-motive* architecture, according to Oosterhuis, is a research program for the design and production of the so-called *Hyperbody*, "a programmable building body that changes its shape and content in real-time."[6] In this vision, buildings are conceptualized as input–output programmable devices that can feed on, process, and transmit information in real-time. They can constantly act and react to different user input and environmental stimuli, as well as to other connected buildings.[7] Besides functional capacities, e-motive architecture would be capable of amplifying experiences and emotions, as well as develop its own autonomy and ability to communicate and interact with its occupants.[8]

Oosterhuis's approach is obviously theoretical; he envisions a future where domestic spaces would be able to transform according to the desires of their inhabitants by means of information technology applications and smart devices, such as microprocessors and sensor/actuator effectors. He argues, however, that such structures and spaces are no more inconceivable or infeasible (as the constructed *Muscle NSA* project, built for the Non-Standard Architectures exhibition at the Centre Pompidou in 2003, demonstrates). Therefore, he thinks, it is now possible to actually inhabit them.[9] In the 1990s, several researchers explored spaces that can respond to human wishes, by means of embedded computer technology. They aspired to implement the so-called intelligent environments (see Chapter 2). At the same time, Oosterhuis's reference to the animate aspects of these spaces, implies insights from fields such as artificial intelligence and artificial life, as well as theoretical and even philosophical questions about agency as a property of inert artifacts, and the boundary between the living and non-living. This book looks into these questions in relation to architecture, taking a long view, from the historical precedent to the present state of the art. It looks precisely at this link between domestic space and ubiquitous computing, from the perspective of a theoretical framework within which the related projects can be addressed.

Several current or past projects – kinetic digitally-driven structures, interactive installations and intelligent environments – make use of ubiquitous computing systems and mechanical apparatuses in order to flexibly adapt to changing conditions, or human needs and desires. The book critically explores their conceptual and technological aspects, along with relevant early precedents – the historical context. Apart from their functional capacities, the book discusses their psychological aspects, drawing from the social studies of technology and the history and theory of artificial intelligence. It also looks into the theory of human–computer interaction and post-cognitivist theories, to suggest the potential implementation of interactive open-ended user-driven environments.

We first take a look at Cybernetics, the scientific discipline that emerged in the 1940s; besides technological innovations, cybernetics suggested a certain way of thinking about the world that involved novel epistemological and ontological questions, paralleled by Cold-War politics and military practices, as well as other related fields, like information theory and general systems theory. Questions about interactivity, information feedback, virtuality, and the relationship between organisms and technology were central to the development of cybernetics, which, as Charlie Gere explains, is the direct predecessor of our current digital culture.[10] It is within this culture that some of the first attempts to create computer technology-enhanced architectural spaces emerged (from the work of Archigram and Cedric Price in Britain to that of Nicholas Negroponte in the US). This historical precedent cannot be dismissed, as the same vision of spaces that would be able to respond to human needs and desires revived in the 1990s, in academic or independent research groups. Assisted by research projects in the field of Ambient Intelligence and Ubiquitous Computing, and the proliferation of wireless computer technologies, this vision became less speculative and theoretical. Built projects appeared in places such as the department of Architecture and Technology at TUDelft and the associated Hyperbody Research Group, the Media Lab at MIT, the Design Research Lab at the Architectural Association, and the Interactive Architecture Workshop at the Bartlett School of Architecture.

But these built projects were far from public awareness and hardly applied to architectural practice, although in the last 15 years several researchers and practitioners publicized their work and set the pace for the wider dissemination of this kind of architecture. Kas Oosterhuis's early books *Architecture Goes Wild* (Rotterdam: 010 Publishers, 2002), *Programmable Architecture* (Milan: L'Arca Edizioni, 2002) and *Hyperbodies: Towards an E-motive Architecture* (Basel: Birkhäuser, 2003) along with his later books *iA#1 – Interactive Architecture* (Heijningen: Jap Sam Books, 2007) co-edited with Xin Xia and *Hyperbody: First Decade of Interactive Architecture* (Heijningen: Jap Sam Books, 2012) co-edited with Henriette Bier, Nimish Biloria, Chris Kievid, Owen Slootweg, and Xin Xia, were paradigmatic of this attempt. Michael Fox and Miles Kemp's book *Interactive Architecture* (New York: Princeton Architectural Press, 2009) was also a concise review of current projects, ranging from kinetic structures to computationally augmented installations and spaces. The authors discussed questions about functional and environmental adaptability, as well as social and psychological implications. Several related articles and projects can also be found in two *Architectural Design* issues, edited by Lucy Bullivant: *4dspace: Interactive Architecture* (London: Wiley 2005), and *4dsocial: Interactive Design Environments* (London: Wiley 2007). These and some other books, along with several journal or conference papers published in the last two decades, constitute the written material that reflects this recent interest in the integration of ubiquitous computing in the built environment. In these texts, keywords like "interactive architecture," "responsive environments" and "smart spaces," are often interchangeable.

A review of these books and research papers showed that a unifying and substantive theoretical framework to contextualize and conceptualize the broad scope of these projects, was missing. It is one of the aims of this book to provide a theoretical backbone, within which a proper discussion can take place, and to fuse this discussion with concepts and theoretical ideas from the social studies of science and technology, post-cognitivist theories, artificial intelligence, human–computer interaction, as well as experimental and common-sense psychology.

Meanwhile, computer engineers made several efforts to create systems in domestic spaces, the so-called *Intelligent Environments*, able to respond to the presence and activities of people, in an adaptive and proactive way, by supporting and enhancing their everyday living habits. This visionary project, called *Ambient Intelligence*, emerged in computer science since the early 1990s, when Mark Weiser suggested the idea of invisible and ubiquitous computing (see Chapter 2). Wireless networks, intelligent agents, user-friendly interfaces, and artificial intelligence, including neural nets and fuzzy logic, were all employed in order to implement this vision.[11] Computer engineers published and disseminated their findings in several conference proceedings and meetings around the world, such as the *International Conference on Intelligent Environments (IE)*, the *International Conference on Pervasive Computing*, the *International Conference on Ubiquitous Computing (UbiComp)* and the *Conference on Internet of Things and Smart Spaces (ruSmart)*. They worked mainly in academic contexts and research groups, such as the Intelligent Inhabited Environments Group at Essex University, the House_n research program at MIT, the Responsive Environments Group at MIT Media Lab, and so on.

These researchers, however, hardly joined the architectural research community that dealt with similar projects, although they would both benefit from sharing and exchanging their views, ideas and findings. Probably the problem was that they lacked a common vocabulary, the fact that architects were unable to acquire in-depth technical knowledge, and also the different goals they aimed at. While engineers were concerned with efficiency and optimization of functionality, architects dealt with formal, aesthetic and experiential aspects. That said, architects kept looking at their projects as another kind of architecture (as Oosterhuis's article "A New Kind of Building" suggested),[12] where form and function were still the main features of architecture, while engineers, funded by private corporations and the industry, were subject to the technocratic feasibility and commercial application of their work.

Currently, however, both communities are interested in the integration of computation in the built environment, and both envision spaces that can be programmed and customized to be able to adapt proactively to the needs and desires of their users. Despite differences, both architects and engineers seem to share a common target: to create environments that can adapt to the people that inhabit them. Therefore, one of our aims in this book is to look at the research outcome of those two communities, and examine their applications and projects from a shared perspective: *adaptation*.

Since the first attempts to introduce information technology and cybernetic concepts in architectural space, adaptation emerged as a potential property of architectural space. In 1969, Andrew Rabeneck proposed the use of cybernetic technologies in buildings, to make them adapt to the future needs of their occupants, and thus aid architects in extending the lifespan of their spaces.[13] In the same year, cybernetics pioneer Gordon Pask, in his article "The Architectural Relevance of Cybernetics," explained how cybernetics could make buildings learn, adapt and respond according to the aims and intentions of their users.[14] Three years later, Charles Eastman, based on earlier ideas by Gordon Pask and Norbert Weiner, suggested an "Adaptive-Conditional Architecture," namely buildings able to self-adjust through information feedback.[15] These are some of the earliest references to the property of adaptation from the perspective of architecture. From the perspective of computer engineering, Norbert Streitz, for instance, explained that the aim of Ambient Intelligence is to create environments capable of responding in a thorough, adaptive, and sometimes proactive way to the presence and activities of people.[16] Thus, the term *adaptation* in the title of the book, points to the idea that architectural space can flexibly adapt to changing conditions and needs (which implies the modernist concept of flexibility that we will address in Chapter 1), but it also assumes another connotation, derived from information theory. It implies a property of what biologists and cyberneticists call adaptive systems, i.e. organisms or mechanisms capable of optimizing their operations by adjusting to changing conditions through feedback.

Cybernetics was a science of historical importance that marked the start of an intellectual revolution, leading to the widespread dissemination of technologies, such as the information and communication networks that organize the infrastructure of the post-industrial world. Chapter 1 examines how cybernetics was applied, either as a conceptual tool or literal technology, in the post-war experimental work of the British Archigram group, the theoretical project of Reyner Banham, the more pragmatic work of Cedric Price and Yona Friedman, as well as the urban vision of Greek architect Takis Zenetos. Cyberneticists tried to synthesize biological organisms with technology, on the basis of cybernetic concepts like information feedback, self-regulation and adaptation. These efforts inspired these avant-garde architects, especially in a period that demanded a shift in thinking about architecture and its relation to communication and information technology. Archigram's iconographic approach to technology-enhanced and user-determined architecture, and Banham's theoretical project, would have seemed rather sketchy for architects like Cedric Price, who even collaborated with Gordon Pask for the implementation of a cybernetic adaptive environment, in his *Fun Palace* megastructure. Yona Friedman's diagrammatic work paralleled Price's cybernetic thinking, but was far from the detailed study of Takis Zenetos's pragmatic project for an *Electronic Urbanism*, which, although never realized, explored ideas that were partly implemented in his other constructed buildings. Although the inspiring designs of those architects were mostly unrealized experiments, similar ideas re-emerged in the 1990s within several research groups or private architectural practices, when the pervasiveness of information technology brought forth, once again, the possibility to synthesize architecture and computer technology, in order to make spaces adaptable to changing needs.

As explained before, in this book we use the term *adaptation* to refer to what may be involved in the interaction between architecture and ubiquitous computing, and to distinguish it from the concepts of flexibility and adaptability. The concept of flexibility in architecture, has two meanings: a determinist flexibility, manifested in the predetermined and mechanistic, closed-system transformations, of the so-called kinetic architectural structures, and a non-determinist, open-ended flexibility (or adaptability), which implies incalculable practices of appropriation by users in space. This attitude to flexibility seems to be linked to the shift towards open-ended adaptive environments, manifested in the designs of post-war visionary architects.

In Chapter 2, we are looking at the legacy of cybernetics, artificial intelligence, as well as the perpetual meaning of *intelligence* in artificial systems. We examine the new AI, robotics and embodied cognition, artificial life and swarm intelligence, in order to closely look at the application of intelligence in the built environment. We further examine the vision of Ambient intelligence (AmI), a research project in computer science for the implementation of *Intelligent Environments*. These are spaces capable of responding proactively to the needs and activities of people, through adaptive systems, ubiquitous computing (neural nets and fuzzy logic supported by networks of intelligent agent-based systems), and user-friendly interfaces. We review a number of respective projects, such as the iDorm, the PlaceLab and PEIS ecology, looking into their functional capacities and social issues. We conclude that, although most of the applications of Ambient intelligence use learning, activity recognition and proactive anticipation mechanisms to adapt to changing human needs and desires, in the manner proposed by the experimental architects discussed in Chapter 1, they essentially respond to already preconditioned preferences, through automated rule-driven services. Interesting counter-examples come from the so-called *end-user driven* systems, namely environments that facilitate user involvement in their functionality.

The introduction of ubiquitous computing in architectural space raises several questions about psychological implications. If physical structures can act, perceive and react by means of sensors and actuators to the activities that take place around them, then people may perceive them as entities that demonstrate some level of intelligence and agency. In Chapter 3, we address both the psychological dimensions of the perception of animacy in inanimate objects, and the culturally defined tendency to construct artificial living machines. We associate this exploration with some examples of computationally augmented architecture, such as Hyperbody's *Muscle Tower II*, and Philip Beesley's *Hylozoic Ground*. The drive to design and construct architecture that can move and interact by means of computational systems, implies the idea of the boundary object, a concept that sociologists of technology have recently explored, to refer to computational objects (or rather quasi-objects) that challenge the boundary between the animate and the inanimate. This idea is part of the history and theory of artificial life, the practices of production of animate objects since the 18th century, and can be further associated to the theoretical criticism of modernity's object–subject or nature–culture divide, and human–machine discontinuities. In light of these observations, we are led to question the concepts and practices pertaining to computationally augmented architecture.

Chapter 4 proposes a theoretical framework to contextualize interaction between users and computationally augmented architecture. We examine the symmetrical and non-symmetrical associations and synergies between people and interactive environments, drawing insights from human–computer interaction (conversational, situated and embodied models of interaction), social and post-cognitivist theories of technology (actor–network theory, distributed cognition and activity theory), and theories of technology that use insights from the phenomenological tradition (the post-phenomenological approach to technology). The chapter starts with the question of interactivity and interactive architecture, addressing, at the same time, its historical precedent in technological and cultural manifestations of the post-war environment, and the emerging cybernetic paradigm. We further pursue the concept in contemporary digital media, and examine its particular meaning within the fields of conversation analysis and human social interaction, in order to explore its potential application in computationally augmented spaces. In light of this discussion, we review contemporary interactive environments and installations, such as *ADA* and *Performative Ecologies*, in order to evaluate their interactive capacities.

In Chapter 5 we look at alternative approaches to traditional Human–Computer Interaction, in order to explore augmented environments that can be flexibly customized by users. We examine how tangible computing applications and end-user driven intelligent environments, like *PiP* (pervasive interactive programming) and e-*Gadgets*, allow users to program and appropriate their environment by "building" their own virtual appliances, according to their indeterminate and changing needs. These environments constitute "ecologies" of interacting artifacts, human users, infrastructure and software, i.e. heterogeneous components that demonstrate open-ended, indeterminate and multiple functionalities. Along these lines, the chapter goes on to suggest the "open potential" environment, an assemblage of "intelligent," programmable, and customizable devices and software for domestic spaces. Its benefit, however, is that it can also act on its own, according to a set of incorporated social motives, which is a condition for conversational and collaborative processes. This approach to design, which deals with interactive domestic infrastructures (smart reprogrammable and customizable objects, devices and appliances), suggests novel understandings of spatiality and a shifting concept of place. It questions the notion of domesticity, and highlights changing patterns of spatial and temporal organization of the household, which influence the rituals of daily life.

The book synthesizes ideas and projects from diverse fields: from architecture and design to the social studies of technology, computer science and human–computer interaction. It explores experimental precedents and the related history, suggesting possible links between past and recent projects, with a particular emphasis on the concept of adaptation. The book also takes an interdisciplinary perspective, in order to construct a coherent conceptual framework that could help researchers and academics assess relevant projects, current tendencies, and work towards future design and implementation. Therefore, it is aimed at students, professionals and academics interested in the link between architecture and digital technology, or more broadly in spaces and artifacts mediated by intelligent systems. It aspires to become a platform for potential debate and critical response, for further research and development, bringing together architects, artists and engineers.

Notes

1 Reyner Banham, "A Home Is not a House," *Art in America* 2 (1965): 109–118. Reprinted in *Rethinking Technology: A Reader in Architectural Theory*, eds. William W. Braham and Jonathan A. Hale (London/New York: Routledge, 2007), 159–166.
2 Nicholas Negroponte, *Soft Architecture Machines* (Cambridge, MA/London: MIT Press, 1975), 133.
3 Kas Oosterhuis is the director of the Hyperbody Research Group and professor of digital design methods at the department of Architectural Engineering and Technology, Delft University of Technology (http://www.hyperbody.nl). See also the *Hyperbody Research Group* website at TUDelft at http://www.bk.tudelft.nl/nl/over-faculteit/afdelingen/architectural-engineering-and-technology/organisatie/hyperbody.
4 The *E-motive House* is a proposal for a speculative programmable structure, able to change content and form by means of sensor and actuator devices, responding to the activities of local or internet users. All servicing functions are placed at the outer ends of the structure to leave the space in-between open and functionally unspecified. The house incorporates modular structural components to transform by shifting its walls, floor or ceiling, in order to adapt to different and changing functional requirements, according to needs and desires (see Chapter 3). See Kas Oosterhuis, "E-motive House," *ONL [Oosterhuis_Lénárd]*, 2006, accessed June 24, 2014, http://www.oosterhuis.nl/quickstart/index.php?id=348 [site discontinued].
5 This is one of a series of speculative structures that can respond to incoming information from local visitors and internet users, by changing their overall shape and content (as projected information in the interior) in real time and in unpredictable ways. The structure would be enveloped by a mesh of pneumatic actuators that can change their length, and thus the overall form of the building, according to the air that is pumped into them. See Kas Oosterhuis, "A New Kind of Building," in *Disappearing Architecture: From Real to Virtual to Quantum*, eds. Georg Flachbart and Peter Weibel (Basel: Birkhäuser, 2005), 106–107.
6 See "E-motive Architecture – Inaugural Speech, Delft University of Technology," accessed June 24, 2014, http://www.bk.tudelft.nl/en/about-faculty/departments/architectural-engineering-and-technology/organisation/hyperbody/research/theory/e-motive-architecture/2001-e-motive-architecture.
7 Ibid.
8 Oosterhuis, "E-motive House."
9 Ibid.
10 Charlie Gere, *Digital Culture* (London: Reaktion Books, 2008), 80–81.
11 Not to mention the fact that intelligent building systems are increasingly and widely embedded now in many buildings around the world, to manage lighting, security, privacy, air-conditioning and temperature systems, while radio frequency identification (RFID) tags are increasingly attached onto products and consumer goods, to enhance control and accounting practices.
12 See Oosterhuis, "A New Kind of Building," 90–115.
13 Andrew Rabeneck, "Cybermation: A Useful Dream," *Architectural Design* 9, (1969): 497–500.
14 Gordon Pask, "The Architectural Relevance of Cybernetics," *Architectural Design* 9, (1969): 494–496.
15 Charles Eastman, "Adaptive-Conditional Architecture," in *Design Participation: Proceedings of the Design Research Society's Conference*, Manchester, September 1971, 51–57 (London: Academy Editions, 1972).
16 See Norbert Streitz's paper: "Designing Interaction for Smart Environments: Ambient Intelligence and the Disappearing Computer," in *The 2nd IET International Conference on Intelligent Environments (IE06)*, vol. 1, Athens, 5–6 July 2006 (London: IET publications, 2006), 3–8.

1
ARCHITECTURE, ADAPTATION AND THE CYBERNETIC PARADIGM

Reyner Banham, Cedric Price, Yona Friedman, Nicholas Negroponte, the members of the Archigram group, as well as the Greek architect Takis Zenetos, were interested in the application of cybernetics in the built environment, as either a conceptual tool or literal technology. They employed cybernetic concepts, like indeterminacy, information feedback, self-regulation and adaptation, in their designs and writings, to envision open-ended user-driven environments. Although their projects were mostly unrealized experiments, similar ideas re-emerged in the 1990s, when the ubiquity of digital devices brought forth, once again, the possibility to synthesize architecture with computer systems, in order to make spaces adaptable to changing needs. This synthesis is what the term *adaptation*, in the title of the book, refers to, and, as we will explain later, is distinguished from the architectural concepts of flexibility and adaptability. Therefore, the term adaptation takes another connotation, derived from biology and cybernetics; it pertains to the function of adaptive systems, namely the capacity of systems to change and improve their performance, by adjusting their configuration and operations in response to environmental information feedback.

Cybernetics – The Cultural and Scientific Context

Towards a New Kind of Machine

In his 1947 lecture "Machine et Organisme" (first published in 1952),[1] the French philosopher and historian of science Georges Canguilhem argued for a reversal of the mechanistic philosophy and biology, which explained organisms in terms of mechanistic principles. Instead he put forth a biological understanding of technology and machines, and located his view in the anthropological interpretation of technology, which regarded tools and machines as extensions of human organs, and owed a lot to ethnographic studies, and the work of anthropologist André Leroi-Gourhan.[2] Thus, Canguilhem attempted to explain the construction of machines in different terms than just being the result and application of scientific activity and theorems, or even the processes of industrialization in the capitalist economy. He proposed an understanding of machines by making reference to the structure and functions of organisms, which, unlike the quantifiable, predictable, calculated and periodical movements of mechanical systems, are characterized by self-regulation, auto-maintenance, reproduction, polyvalence and openness to potentiality. Canguilhem traced this view in a new emerging discipline in the US at the time, called "bionics."[3]

These ideas anticipated a research agenda put forth by the new science of cybernetics, founded by Norbert Wiener in the 1940s. A year after Canguilhem's lecture, Wiener defined cybernetics as the "science of control and communication in the animal and the machine, organisms and mechanisms."[4] In his book *The Human Use of Human Beings*, he discussed the idea of information feedback and self-regulation as the bases for understanding biological, machinic and social processes. Thus, Wiener suggested that it would be possible to create a new kind of machine, namely an environmentally responsive machine that would put the principles of the living to work in the world of the non-living.[5]

Until the mid-20th century, scientists considered the idea of constructing machines that are more complex than man to be a mere dream or a cultural fantasy. But in the 1940s and 1950s many scientists started to think of novel ways to examine the complexity of intelligent behavior in men, animals and machines, as well as the transmission and reception of messages within a feedback system, either biological or technological. Synthesizing information theory, communication theory and the theory of machine control, they established cybernetics and formulated its basic principles in the so-called Macy conferences. Mathematicians, physiologists and engineers, such as Norbert Wiener, John von Neumann, Claude Shannon, Warren McCulloch, Margaret Mead, Gregory Bateson and others, met in these events from 1942 until 1954, to discuss their work and its interdisciplinary applications. Cybernetics marked the start of an intellectual revolution out of which scientific fields such as cognitive science and artificial intelligence emerged, as well as widespread technologies, such as the information and communication networks that organize the infrastructure of the post-industrial world.

The proponents of cybernetics talked about machines in terms of living organisms and about living organisms in terms of machines. Moreover, most of the electromechanical devices that they constructed to demonstrate the principles of cybernetics in action, manifested an apparent intelligent and adaptive behavior, analogous to living organisms. Ross Ashby re-contextualized cybernetics, not only within information theory, but also within dynamical systems theory, arguing that cybernetics constituted a theory of every possible machine. It was a framework within which all machines, constructed either by man or nature, could be classified according to what they can do and how they can possibly function, and not what they are. Thus, the aim of cyberneticists was to bridge, ontologically, the gap between the organic and the inorganic, the natural and the artificial, organisms and technical objects, acknowledging the difference in the complexity that comes in their respective organization. They proposed, not a model or a simulation of living organisms, but a new kind of machine, which would transcend the opposition between the two poles. In this context, the complexity of life would not be attributed to some untold secret power, as in vitalism, nor would it be reduced to a determinist, linear, mechanical causality, as in mechanistic philosophy and the Cartesian conceptualization of animals as complex machines.[6] Transcending the debate about the difference between living organisms and machines beyond vitalistic terms, Wiener suggested that their equation, from a cybernetics standpoint, would arise from the fact that both entities could be regarded as self-regulating machines, able to control (or better reduce) entropy through feedback; that is, the capacity to regulate future behavior from previous operations and eventually learn from their experience. These machines had actuators to perform specific tasks, and sensors that allowed them to record external conditions as much as their own operation, and return this information as input for their next action in a continuous self-regulatory circuit.[7]

Self-regulation characterized the systems of the first cybernetic wave (1945–1960). It was a property of "homeostasis" – physiologist Walter Cannon coined the term in 1929 – namely the ability of animal bodies to maintain a stable energy status, and to minimize energy loss in relation to environmental changes. Respectively, cyberneticists considered the homeostatic cybernetic model to be able to minimize information loss in the system, in favor of control and stability, through "negative feedback." A homeostatic machine would resist the tendency towards disorder, due to environmental disturbances, i.e. "noise," eliminating possible unexpected behaviors and state changes (Figure 1.1). Such a system would be predictable, because it would use feedback loops to constantly follow its target, and maintain the stability of its state of operation, correcting all possible destabilizing activities.

Homeostasis and Architecture

In 1972, Charles Eastman suggested that architecture could be conceived as a feedback system self-adjusting space to fit user needs in a dynamic stability, such as that of a boat constantly correcting its course against environmental disturbances.[8] Similarly, in 1975 Nicholas Negroponte, discussing machine controlled environments, regarded greenhouses with opening–closing glass roofs as examples of homeostatic buildings, i.e. buildings working as thermostat mechanisms, able to control, decode, interpret and encode information about the comfort level of plants.[9] Michael Fox recently described such a system for kinetic architectural structures, through a diagram of what he called "responsive

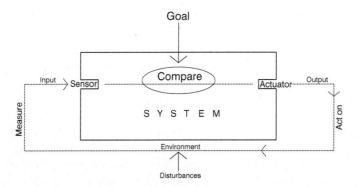

FIGURE 1.1 First-order cybernetic system

indirect control"; a central computer program receives environmental sensor input and sends an optimal instruction through regulatory feedback to actuators to move a single architectonic element.[10]

Thus, a homeostatic system in buildings would be able to constantly adjust the state of the internal environment through feedback, according to the given needs of its inhabitants or the overall goals of the system (Figure 1.2). For instance, it would include environmentally responsive building skins, capable of regulating and correcting their transparency and permeability (by means of kinetic blinds or mobile roof panels) through a closed feedback loop, registering environmental information (sunlight intensity, humidity levels, wind speed) towards a predetermined goal (for instance stable light intensity). A homeostatic architecture, however, would involve predictable behaviors functioning towards

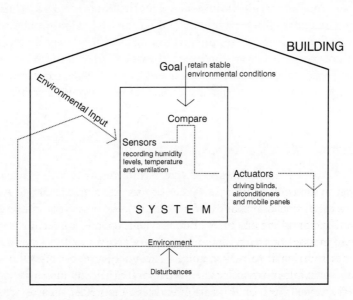

FIGURE 1.2 Homeostatic system in buildings

constant elimination of unwanted "disturbances." Thus, it would only work in a context where environmental fluctuations and operational changes are predefined and known in advance.

This is well illustrated in an early experimental project built by Nicholas Negroponte and MIT's Architecture Machine Group, called SEEK (aka Blocksworld), which took issue with both first-order and, as we will see in a moment, second-order cybernetics. SEEK was a computer-controlled installation originally shown at the "Software" exhibition at the Jewish Museum in New York 1970, curated by Jack Burnham. The installation consisted of 500 two-inch metal cubes on a rectilinear three-dimensional arrangement within a plexiglas box populated by a number of gerbils, constantly relocating the cubes and changing the overall set-up. As a response to these "disturbing" actions, a gantry-type robotic arm with an electromagnetic end-effector (fed by a video camera that registered the changes), would constantly rearrange the cubes; those that were slightly dislocated would be realigned according to the computational model of the set-up – the programmed three-dimensional configuration stored in its computer memory. But if the cubes were significantly dislocated by the gerbils, the system would move them in their new place on the assumption that the gerbils "wanted" them there.[11]

SEEK was an experiment in the symbiotic relationship between living organisms and machines, as well as an exploration into the capacity of the machine to deal with the mismatch between its computational model of the structure and its real-world configuration. More importantly, considered as a model of urban and architectural design, the overall set-up would be either reconstructed back to its initial model or to a novel configuration. If the gerbils' erratic and unpredictable behavior is the analogue of the system's "noise" that the negative feedback eliminates through "corrective" operations (redistribution of the cubes by mechanical means), then the project can be regarded as a model of a first-order cybernetics approach to architecture. However, when the system relocates the cubes, reorganizing them according to the indeterminate actions of the gerbils (the "wishes" of its occupants), it amplifies, rather than eliminates, "noise," which pertains to a second-order cybernetics model.

Second-order Cybernetics and Autopoiesis

Heinz von Foerster formulated second-order cybernetics between 1968 and 1975 by exploring the constitutive role of the observer in the formation of systems, as well as the virtues of positive feedback, that is, the process of multiplication and amplification of the amount of information. In a positive feedback system, a distorted message or a perturbation would reinforce the system's organization, sometimes spurring it to self-organize (and not decompose it), an idea that anticipated later self-organization theories.[12] Sources of inspiration for the second cybernetic model were findings in other scientific fields, such as chemistry and genetics, although it reached maturity when Humberto Maturana and

Francisco Varela published their book *Autopoiesis and Cognition: The Realization of the Living* (1980). In the autopoietic model, environmental stimuli trigger the internal organization of the system and its components. This operation preserves the identity of the system, which is determined by that organization. While the homeostatic model uses information to retain order in the system by interacting with the environment, the autopoietic system is autonomous and closed, constantly reproducing and regenerating the components that constitute it. Because of its emphasis on the process, autopoietic theory is more useful in the analysis of social, not technological systems and machines; the way Varela articulated autopoietic theory facilitated its generalized application in systems, either biological, urban or social.[13] But to deal with machinic processes, Maturana and Varela proposed another more suitable concept, *allopoiesis*, namely the process by which the system produces something other than its constitutive components (as, for instance, a car factory produces something other than the industrial robots of its assembly line).

Second-order cybernetics provided the framework for modeling complex control systems, like GPS-guided automated steering systems, or for the design of software and service systems that deal with user motivation and manage the hierarchical organization of goals (Figure 1.3). But, as we shall see in later chapters, second-order cybernetics and autopoietic theory led to the understanding and modeling of complex and adaptive processes, such as self-organization, emergence, learning and conversational interaction in systems

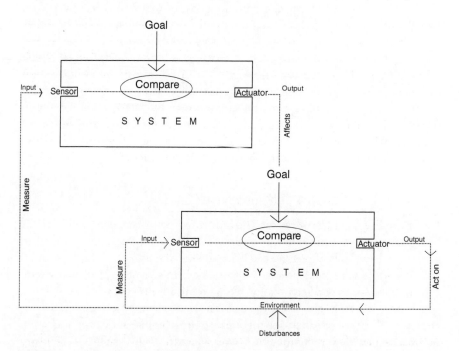

FIGURE 1.3 Second-order cybernetic system

– either artificial, social or biological. As we shall see in the next chapter, researchers working in the field of so-called *Intelligent Environments* since the beginning of the 1990s in several academic (and non-academic) research labs, have explored the potential of those further developments stemming from second-order cybernetics. Furthermore, since the 1990s, architects Stephen Gage, Usman Haque, Ruairi Glynn, Ted Krueger, Gillian Hunt, and Ranulph Glanville, have taken interest in the possible implementation of experimental cybernetics-driven environments, beyond the principles of homeostasis, and towards more interactive models, expounding on the ideas and work of early cyberneticists like Gordon Pask, as well as the architectural avant-garde of the 1960s.[14]

Before we come to these recent approaches, in the following section we will look at the work of architects and groups of the avant-garde who, before Negroponte's SEEK, demonstrated a radical understanding of the role of information technology and cybernetics in architecture. We will see that although their experimental work engaged mechanical and kinetic flexibility techniques (and therefore determinist), they expressed a vision for truly adaptable and open-ended user-determined environments.

The Architectural Avant-garde and Cybernetics

Architecture, Flexibility and Adaptability

Throughout the 20th century, architects sought ideas, techniques and strategies to make buildings, and in particular domestic spaces, flexible. This meant the capacity to adjust to changing needs, and respond to demographic, economic, and environmental conditions. The most common and well-known application of flexibility in modernism involved sliding partitions and foldable furniture. This type, what Adrian Forty calls *flexibility by technical means*,[15] was applied in modernist buildings like Gerrit Rietveld's Schröder Huis,[16] Johannes Van den Broek's Woningenkomplex Vroesenlaan,[17] or Le Corbusier's Maisons Loucheur (Figure 1.4).[18] It was a tactic to represent the dynamism of modern

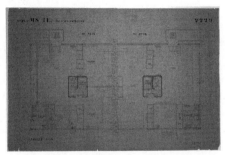

FIGURE 1.4 Le Corbusier "Maison Loucheur, 1929, Plan FLC 18238" © FLC / ADAGP, Paris / OSDEETE, Athens 2015

life in housing, aligned with the notion of progress and mechanized technology in the cultural context of modernity. Most importantly it is a predetermined, or "hard" flexibility, as Tatjana Schneider and Jeremy Till call it, because it involves technical elements that specifically determine the way people will use their spaces.[19] Therefore, as Forty explains, it was simply an extension of functionalism, because it gave architects the illusion that they can sustain and extend their control on buildings even after the period of their real responsibility, the design stage.[20]

On the other hand, although the practices for controlling and ordering domestic space were part of modernist functionalist design strategies,[21] throughout the 20th century flexibility acquired connotations beyond its determinist sense. This became clear in the post-war period in particular, when architects and theorists took it up as a concept to criticize determinist functionalism and deal with the uncertainty and rapid development in the Western world.[22] In fact, as Forty explains, flexibility in modernist discourse

> [w]as a way of dealing with the contradiction that arose between the expectation… that the architect's ultimate concern in designing buildings was with their human use and occupation, and the reality that the architect's involvement in a building ceased at the very moment that occupation began.[23]

Therefore, in the 20th century, other proposals addressed tactics for a certain indeterminacy, allowing users to adjust their spaces according to their needs beyond the architect's control. What Schneider and Till call "soft" flexibility,[24] or rather *adaptability*, was the capacity for different social uses through the organization and connection of rooms – not through physical alterations and technical means (such as sliding partitions, extensions of rooms, folding furniture etc.).[25] This would usually point to the idea of the *open plan*,[26] namely a plan with functionally unspecified rooms, generally without physical changes (as in Bruno Taut's Hufeisensiedlung housing estate), and sometimes with some *redundancy*.[27] The term also covers what Hertzberger later called *polyvalence*, although his emphasis was on the way spatial form was mentally interpreted by users to enable a (finite) range of possible uses in functionally unspecified spaces (as in Hertzberger's Diagoon houses).[28]

Thus, as Forty suggests, flexibility in architectural design and discourse was employed on the one hand to extend functionalism and on the other to resist it. In the former instance, flexibility was a property of buildings (implemented by means of mechanical changeability of the plan); in the latter it was a property of spaces and the use that they are put to. Following Henri Lefebvre's theory of space and his critique on functionalism, Forty goes on to argue that in this latter case flexibility acquired its political connotations.[29]

Lefebvre's theory of space is mainly drawn from his book *The Production of Space* (first published in French in 1974), which contains one of the most radical and comprehensive critiques of space, and its production in relation to the structures of power and domination in modernity. In Lefebvre's theory, space is foremost social, rather than a distinct category of architecture, or a quantifiable thing, which architects can claim to have sole authority upon. Lefebvre treats space as a conceptual triad; it intertwines the perceived space (involving spatial practice, the production of society's material spaces), the conceived space (its representations by planners, urbanists, scientists, social engineers etc.) and the space as directly lived and inhabited by the body (what Lefebvre calls representational space).[30] Lefebvre sees functionalism as a characteristic of the space of modern capitalist societies, which is a historically specific form of spatial production that specifies and devalues use: "Functionalism stresses function to the point where, because its function has a specially assigned place within dominated space, the very possibility of multifunctionality is eliminated."[31] Lefebvre exposed the problematic distinction that is prominent in modern capitalist society, between space as an abstract construct, aligned to strategies of domination and homogenization, and the *lived* space of users' everyday practice, favoring qualitative difference, imagination and appropriation: "The user's space is lived – not represented (or conceived). When compared with the abstract space of the experts (architects, urbanists, planners), the space of the everyday activities of users is a concrete one, which is to say, subjective."[32]

This latter case affects our conceptualization of users. Users are not abstract, undifferentiated and homogeneous entities, as defined in the context of orthodox modernism and the post-war state-funded housing programs in western countries. Instead, in many instances, the term "user" acquired a positive connotation, meaning an active creative agent.[33] In this sense, users are not simply occupants; they can appropriate space, determine its manner of use at will, creatively re-interpret it or even misuse it.[34] This space of everyday practice, which is not regulated by functionalist design, pertains to another type of flexibility, what Forty calls *flexibility as a political strategy*,[35] i.e. a strategy *against* architecture, functionalism and determinist design techniques. In Michel de Certeau's theory of everyday life, this strategy points to what de Certeau calls "tactics," i.e. practices and "treatments of space" by which people manipulate, transform and at the same time resist pre-established structures of imposed order.[36]

Of course users are always creative and unpredictable; most buildings and spaces, especially residential, are transformed and appropriated no matter how "open" or, conversely, determinate the plan is, and despite the initial intentions of their architects and designers.[37] What we are exploring in this book, however, is the possibility to empower users to adjust their environments in unpredictable and individual ways, using the capacities of ubiquitous computing. In the following we will look at the post-war avant-garde architects that pursued this idea, and took inspiration from cybernetic thinking, to apply computational technology in the built environment.

The Socio-cultural Context of the Architectural Avant-garde

One of the central aims of the architectural avant-garde of the 1960s was to create "open-ended" spaces able to respond to indeterminacy, by empowering users to actively determine their environments. Although their designs depicted "hard" flexibility techniques (mechanical and kinetic systems), they also employed communication technologies and cybernetic systems for constant adaptation to changing and uncertain demands. Contrary to the functionalist attributes of stability, normalization and control, their projects assumed unstable, changing configurations through cybernetic feedback. The very advent of cybernetics was accompanied by a certain shift in the definition of the machine and its relation to organisms, which led to a biotechnical tendency to naturalize machines and thereby buildings. Architects did not describe buildings through functionalist biological metaphors as in 19th-century theory, but rather as conceptualizations of literal environmentally and socially adaptive systems. In this case, flexibility was neither a property of buildings (as in a functionalist perspective), neither of space (which opposed functionalism); these architects linked the idea of flexibility to the potential of architecture to operate as an interactive machine, an adaptive system informed by user input. Thus, within the unpredictable development of the post-war city (including the shift to a post-industrial production system and the uncertainty created by the constant demographic, social and economic changes), cybernetics was used either as an inspirational tool or as literal technology to model self-regulation and systematize uncertainty and changeability. But, as we shall see, buildings were not always imagined as literal cybernetic machines. The analogy between architecture and cybernetic systems was sometimes indirect, metaphorical or sketchy (as in the work of Dutch Structuralists, the Japanese Metabolists and the British Archigram group), and other times pragmatic and literal (as in the work of Cedric Price, Nicholas Negroponte and Yona Friedman).

Of course, architectural thinking and practice interfaced not only with cybernetic ideas in the post-war period, but also with the wider cultural production. This was a post-war environment of growth and optimism; people thought that technology and science would lead the way to social emancipation, while the market of mass-produced and mass-consumed expendable products should be sufficiently flexible and mobile to adapt to the changing desires and multiple lifestyles of people. At the same time, scientific and philosophical discourse was presenting a new paradigm of the world; this was no longer characterized by the mechanistic model of the 18th and 19th centuries, but by Einstein's relativity theory, Werner Heisenberg's "uncertainty principle" in quantum mechanics and Karl Popper's ideological pluralism and his attack on socio-political determinism. This paradigm was also informed by the emerging fields of Information Theory, General Systems Theory, Structuralism, and Artificial Intelligence. Most importantly it reflected cybernetic thinking and ideas such as interactivity, information feedback, and adaptation.[38] These cybernetic concepts not only surfaced in the military funded computational applications of the Cold-War US, but were also concerns of the 1950s and 1960s avant-garde, such the Fluxus art movement, John Cage, the French Lettrists and the Situationist movement. Their writings, practices and performances implicated concepts like participation, improvisation, play, choice, feedback, "noise," information dissemination, connectivity, incompleteness, indeterminacy and open-endedness. In 1950s Britain, the avant-garde Independent Group, inspired by the American culture of prosperity and the belief in the benefits of advanced technology and science, adopted cybernetic and information theory ideas along with non-Aristotelian thinking against British parochialism and post-war austerity. In 1951, the Group member Richard Hamilton put together an art exhibition called *Growth and Form* at the Institute of Contemporary Art in London. Its curators were inspired by D'Arcy Wentworth Thompson's book *On Growth and Form* (1917), and were concerned with expendable aesthetics, images of flexibility, disorder, growth, flux, and biological and technological change, which were also central concerns for the Archigram group.

Indeterminacy in the Work of Archigram

The Archigram group comprised a number of British architects, who mostly worked with speculative paper projects during the 1960s, and disseminated their ideas through an irregularly published magazine bearing the name of the group. Dissatisfied with the architectural status quo, the group took a radical interest in technology-driven architecture and the opportunities of consumer culture, drawing academic credibility for their proposals from historian Reyner Banham's theoretical work.[39] The group envisaged an indeterminate architecture,[40] namely open-ended environments, subject to change, flexible and adaptive to users' needs that would be able to accommodate the accelerating social heterogeneity and subvert the practices of homogenization of modernist design. In their article "Exchange and Response," the group members described the kind of architecture they were trying to achieve:

> An active architecture – and this is really what we are about – attempts to sharpen to the maximum its power of response and ability to respond to as many reasonable potentials as possible. If only we can get to an architecture that really responded to human wish as it occurred then we would be getting somewhere.[41]

These ideas were initially depicted in a series of drawings of urban megastructures, such as those of the *Plug-In City*, which expounded on the schemes of the Futurists and the members of the Independent Group, Alison and Peter Smithson (later members of Team 10). Similar images of indeterminate expandable megastructures could be traced within the work of the concurrent Japanese Metabolist movement and their drawings of capsule towers, although they were driven by an ideology quite different from that of Archigram.[42] According

20 Architecture, Adaptation and the Cybernetic Paradigm

to Alan Colquhoun, such megastructural forms were primarily concerned with self-regulation, a property of cybernetic machines which would allow buildings to adapt to changing human desires. But as he explains, this was more profound in the megastructures of Cedric Price, Yona Friedman and Constant.[43]

However, the group's interest in collectivity and social unity (rather than heterogeneity) that was reflected in the image of the megastructure, gradually faded. It was replaced by the individuality expressed in smaller and more intimate projects, such as the *Living Pod, Auto-Environment, Living_1990* and *Cushicle*. Inspired by pop and consumer culture, science fiction images, and space technology, these projects suggested the conceptualization of buildings as mass-produced, expendable and interchangeable kits-of-parts, accommodating freedom and user choice. In Archigram 8 the group asserted:

> It is now reasonable to treat buildings as consumer products, and the real justification of consumer products is that they are the direct expression of a freedom to choose. So we are back again to the other notions of determinacy and indeterminacy and change and choice.[44]

The Archigram members made use of images of geodesic skins (inspired by Fuller's geodesic domes), inflatable and lightweight structures, as well as mechanical, hydraulic and electronic infrastructure, to depict flexible, mobile and adaptable environments. At the same time the group examined the potential of electronic equipment, robotics and cybernetic interfaces. For instance, the project *Living 1990* (1967), was implemented in the form of an experimental installation about future dwellings at the Harrods megastore in London, to "demonstrate how computer technology and concepts of expendability and personal leisure might influence the form of future homes" (Figure 1.5).[45] The group imagined it as a plug-in unit installed on a larger megastructure, comprising exchangeable and expendable accessories, inflatable beds and seats,

FIGURE 1.5 *House 1990*, View of Installation, 1967. © Archigram 1967.

multi-functional robots, traveling "chair-cars," radio and TV with on-demand programs, as well as adjustable spatial boundaries (floor, walls and ceiling) Users, conceptualized as consumers, could choose the layout of space, and the location, color or softness of its ceiling, floor and walls according to their individual physiological and psychological needs:

> The push of a button or a spoken command, a bat of an eyelid will set these transformations in motion – providing what you want where and when you need it. Each member of a family will choose what they want – the shape and layout of their spaces, their activities or what have you.[46]

The Archigram group envisioned flexible and functionally indeterminate spaces. This reflected their preference for biological characteristics in architecture, as was evident in their writings. For instance, in the 8th Archigram issue they used the term "Metamorphosis" to refer to a transient state of things conceived as alive and constantly evolving.[47] By rejecting conceptual boundaries between organic and inorganic systems, and by designing environments that synthesized biological and technological systems, they wanted to transfer the dynamic aspects of the natural body to the built environment. Following the cybernetics agenda, where both natural and artificial systems shared exchangeable properties, they envisioned environments as "inorganic" extensions of "organic" users, as depicted in Mike Webb's *Suitaloon and Cushicle* (1966) (Figure 1.6). In this project, the human body seemed to extend its boundaries by means of a PVC membrane that enclosed it, acting as a sort of external second skin. Architecture was reduced to clothing, thus overriding the dichotomy between the technological and the natural. Echoing Marshall McLuhan's idea of architecture as an organic extension of the human body and regulator of environmental perception,[48] such personalized environments implemented the idea of a responsive and interconnected cybernetic feedback system.

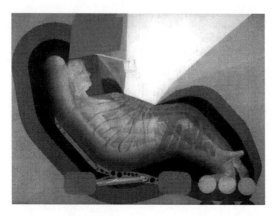

FIGURE 1.6 *Suitaloon & Cushicle* in Reclining Position, with Pat and Tide Logo, Michael Webb, 1966. © Archigram 1966.

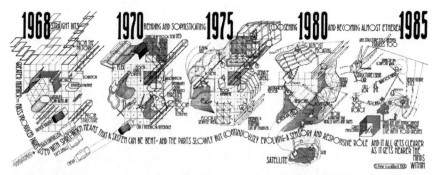

FIGURE 1.7 Control and Choice, *Timeline*, Peter Cook, 1967. © Archigram 1967.

Architecture Without Architecture

The Archigram group gradually attempted to dematerialize architecture, by abandoning both large megastructural schemes and small-scale enclosures inspired by automobiles, trailer homes and inflatables. In Archigram 7 (1966), Peter Cook suggested that "There may be no buildings at all in Archigram 8,"[49] whereas in Archigram 8 (1968), an article titled *The Nomad*, concluded with the claim that "'place' exists only in the mind."[50] This statement pointed to a shift from the "hardware" to the "software," a distinction proposed by the group in the same issue, to imply the shift from the tangible aspects of architecture to its immaterial entities – the system, the transmitted message or program.[51] On the larger regional scale this shift was clearly seen in their *Instant City* project of 1969, a touring air-traveling assemblage of instant enclosures and audio-visual equipment, enhanced by communication networks, and aiming to intensify social and cultural life in towns. But from a more private domestic perspective, Peter Cook's drawing *Timeline* (1967), which was part of the larger *Control and Choice Dwelling* housing project, depicted a formal evolution of a prefabricated living room in a 17-year timeline, as its walls gradually turned into televisual membranes and sensor networks (Figure 1.7). Later projects emphasized even more this theme, suggesting a shift towards a completely virtual architecture, as in the *Holographic Scene Setter* (1969) of Instant City, and *Room of 1000 Delights* (1970), both included in the Archigram 9 issue. In these projects, technological devices would set the conditions for people to fulfill and realize their desires and dreams.[52]

Thus, a cybernetics-inspired mentality led the group to blur the difference between architecture and information; architecture was no more about artifacts, objects and hardware, but rather human–machine interfaces, reduced to the software, the programs and networks that make up the environments that people inhabit.[53] As Hadas Steiner argues, architecture in the work of Archigram turned into an assemblage of complex and indeterminate systems, managing and transmitting immaterial entities of information and energy. The group posed fundamental questions about the nature of architecture, and challenged the mechanistic model of technology that was largely prevalent in modernist architectural theory and practice:

The articulation of buildings as transmitters for intangible entities as transient as energy transfers and information relays raised fundamental questions for the nature of architecture. By challenging the machine-based model of technology that had defined modernist architectural theory and production, the *Archigram* began to represent that which the mind resisted: the dissolution of the artefact – the very concept of an object – into a landscape of complex and indeterminate systems.[54]

The group's obsession with immaterial interfaces, cybernetic feedback circuits, and minimization of hardware, may suggest their attempt to shift away from a functionalist, determinist paradigm of flexibility; yet, we could argue that their exaggerated depiction of technical details and inflexible, mechanical systems, led to rigid, inflexible forms.[55] Therefore, the possibility of architecture without architecture that they suggested, and the structuring of experience without physical constructions, is questioned; ephemeral and immaterial entities, networks, events, movements and information flows, demanded material means and permanent elements to support habitation.[56]

On the other hand, the group's design language, which drew its visual vocabulary from the latest technologies, industrial products, infrastructure, space and underwater hardware, as well as consumerist icons and pop images, was an opportunity to challenge the disciplinary limits and norms of architecture as well as its entanglement with social and political institutions.[57] Thus, while the group members managed to represent, and even symbolize, a cybernetics-driven architecture, their project was largely about exploring the form of this architecture. At the same time, their understanding and use of cybernetic technology was sketchy.

Reyner Banham: The Environment Bubble

Another link between architecture and cybernetics can be traced in Reyner Banham's thinking and work. Like the Archigram members, Banham consistently questioned the conventions and limits of the discipline of architecture. Outdoor living practices (such as those enabled by the trailer home and the drive-in cinema), and the development of domestic services and appliances in the US, gave him the grounds to redefine the concept of the house. Arguing for the nomadic dwelling, such as Buckminster Fuller's Dymaxion House, he proposed the "un-house," an autonomous, adaptable, mobile unit, whose "threat or promise," in Banham's view, was most clearly demonstrated in Philip Johnson's Glass House.[58] Banham's project *Transportable Standard-of-living Package*,[59] a deployable unit of electromechanical domestic services placed at the center of a PVC capsule, the *Environment Bubble*, as well as the project *Anatomy of a Dwelling*, all depicted in Banham's essay "A Home is not a House" (originally published in *Art in America* journal),[60] suggested a shift in architectural priorities; the house and its materiality, its form and envelope,

FIGURE 1.8 François Dallegret, The Environment-Bubble: Transparent Plastic Bubble Dome Inflated by Air-Conditioning output, from "A Home Is Not a House," Reyner Banham, *Art in America* (April 1965). © 1965 François Dallegret.

were quite literally nearing invisibility (Figures 1.8 and 1.9). They anticipated Banham's later book *The Architecture of the Well-Tempered Environment* (1969), where he proposed an alternative architectural history concerned with the environmental control systems in buildings – mechanical ventilation, lighting, and air-conditioning – rather than typology and form. Thus, Banham was able to challenge the very definition of architecture; instead of monumental and permanent structures, he favored responsive, servicing and thus "fit environments for human activities."[61]

In Banham's radical view of technology-enhanced architecture, what was left was the hearth of the house, namely the electromechanical core of services that control domestic life and environment. This is a modern and more efficient version of what he considered to be the second way human beings developed to control their environment, the campfire:

> Man started with two basic ways of controlling environment: one by avoiding the issue and hiding under a rock, tree, tent, or roof (this led ultimately to architecture as we know it) and the other by actually interfering with the local meteorology, usually by means of a campfire, which, in a more polished form, might lead to the kind of situation now under discussion. Unlike the living space trapped with our forebears under a rock or roof, the space around a campfire has many unique qualities which architecture cannot hope to equal, above all, its freedom and variability.[62]

One of the benefits of this second way of controlling the environment, according to Banham, is the variability and freedom offered by an energy

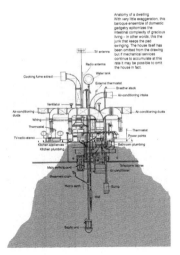

FIGURE 1.9 François Dallegret, Anatomy of a Dwelling, from "A Home Is Not a House," Reyner Banham, *Art in America* (April 1965). © 1965 François Dallegret.

source, i.e. a contemporary air-conditioner, which regulates room temperature according to human needs and desires. Banham's *Environment Bubble*, a speculative proposal for such a cybernetic feedback mechanism, would use responsive technologies to dynamically engage with the local conditions, adapting to environmental changes. Its membrane could supposedly inflate at will and cater for rain, temperature and wind, blowing down an "air-curtain" of either warmed or cooled conditioned air, where and when needed. This would be controlled by light and weather sensors, such as the new invention of the time, the weathervane.[63] Thus, as environmental distractions could be managed by means of adaptive behavior and information feedback, the *Environment Bubble* seems to implement a first-order cybernetic system of self-regulation and "noise" elimination.

But unlike Banham's theoretical approach to the link between cybernetics and architecture, and Archigram's rather sketchy use of technology in design, Cedric Price's vision involved a more pragmatic version of cybernetics-driven architecture.[64] A difference between Banham's and Price's version of environmental control can be traced in the difference between "controlled" and "controllable" or "responsive" environments respectively, as Nigel Whiteley has explained. While the former involves a limited range of choices, the latter is fully responsive to a range of needs or desires, facilitating indeterminate and open-ended situations.[65] As we shall see, Price wanted to seriously incorporate a second-order cybernetics system in his *Fun Palace* project. In particular, he wanted the project to use computational systems to constantly adapt to changing social or environmental conditions, and to manage indeterminacy in both program and form.

Cedric Price's Fun Palace

The Fun Palace (1961–1974), conceived by British architect Cedric Price and theater director Joan Littlewood (founder of the improvisational left-wing Theater Workshop), was a design proposal for a creative, educational and entertaining megastructure, able to foster collective practices, participation and social interaction. The idea emerged in the context of the wider socio-political changes in post-war Britain, and within the optimistic vision for a new post-industrialist society, characterized by automation, and more free time for the working class. Price and Littlewood regarded it as a new idea to shelter the constructive use of free time, and enhance practices of collective improvisation for the production of spatial form. Sited in Mill Meads at the metropolitan area of London, the Fun Palace would be able to motivate different social groups to appropriate and use its infrastructure according to their needs. This would be assisted by means of lightweight transformable components, and the latest information and communication technologies, game theory and a cybernetic system, developed by cybernetics pioneer Gordon Pask. The conceptual framework within which Price designed this transformable and constantly adaptive structure, was characterized by his personal view that buildings were catalysts for social interaction. It also included situationist ideas[66] and Littlewood's vision for an open theatrical space, a sort of university of the streets and a laboratory of active entertainment and performance.

The Fun Palace would be a huge modular structure organized in three rows: a central aisle, housing the collective activities (cinema and theatre), and two laterals, for human activities and services (restaurant, bar, playgrounds, labs) (Figure 1.10). Price emphasized the flowing and flexible character of the building's interior; corridors, rotating escalators and openings would be organized in such a way as to enhance the continuous flow of visitors, while mobile, temporary and inflatable components (walls and rooms), as well as moveable floors, would allow space to stay free of boundaries, open and indeterminate. Adjustable, lightweight suspended spaces would be able to travel around, by means of a mobile crane running along its roof, transforming the interior layout according to changing circumstances. Its adjustable roof blinds would protect from the rain, while steam and warm air barriers, as in Banham's *Environment Bubble*, would replace external walls (Figure 1.11). This emphasis on constant movement and flow, demonstrated Price's understanding of buildings as impermanent structures. He thought that when buildings turned obsolete they should either change location or be dismantled.

Flexibility in the Fun Palace would be realized by means of all flexibility strategies discussed above. The redundancy of the space of the building, as well as its open plan, could allow various events to take place, while different spaces could acquire exchangeable functions; at the same time, technical means and mobile components could change its layout and internal activities at will. Therefore, unlike Hertzberger's polyvalence, which meant that rooms had

Architecture, Adaptation and the Cybernetic Paradigm 27

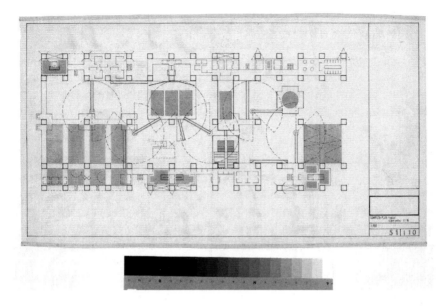

FIGURE 1.10 *Fun Palace*: typical plan, Cedric Price, 1963, ink, screentone appliqué and graphite on translucent paper, 38.4 × 69.4 cm. Cedric Price Fonds, Collection Centre Canadien d'Architecture/Canadian Centre for Architecture, Montréal.

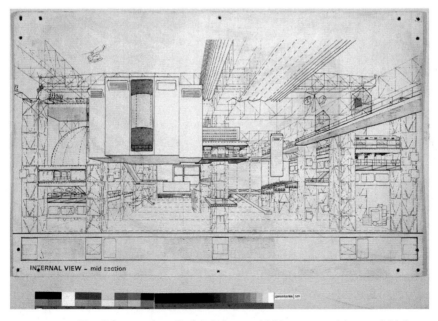

FIGURE 1.11 *Fun Palace*: section, Cedric Price, ca. 1964, reprographic copy laid down on hardboard, 75.8 × 121.4 cm. Cedric Price Fonds, Collection Centre Canadien d'Architecture/Canadian Centre for Architecture, Montréal.

exchangeable functions within a stable form, flexibility in the Fun Palace was implemented by means of a transformable and indeterminate system in terms of *both program and form*, space and structure. At the same time, it involved political connotations, which points to Forty's third type of flexibility, that is, flexibility as a political strategy. By contrast to the Centre Pompidou in Paris, which was inspired by the Fun Palace and is managed centrally and institutionally, Price's project was open for appropriation by its users (the working class of east London), who could both rearrange space and program according to their needs and desires.[67]

According to Stanley Mathews, the Fun Palace reflected a new idea; that architecture was a changeable and transient environment, rather than a symbol of permanence and monumental values. It was an assemblage of temporary, improvisational and interactive systems, able to accommodate sufficient services for social events, adapted to the changing socio-economic conditions and the new knowledge economy of post-war Britain.[68] For Price, architecture should be an open process, an activity involving time, change and indeterminacy, rather than a static form.[69] Like Archigram's and Banham's work, the Fun Palace challenged the very identity of architecture "for it was not really a 'building' at all, but rather a matrix that enclosed an interactive machine."[70] But unlike the Archigram members, who emphasized the visual, Price eliminated the formal attributes in the drawings of the Fun Palace. He replaced representation and symbolic visual language with diagrams that described processes, the active and dynamic aspects of architecture in use.[71]

These processes were properties of not only the technical articulation of the structure but also of the technological systems. In his attempt to model and systematize programmatic uncertainty, in the unstable socio-economic environment of post-war Britain, Price looked at the emerging fields of information technology and cybernetics from a perspective similar to that of Archigram, to create a new kind of adaptive architecture able to constantly regulate its form.[72] But unlike Archigram, technology in the Fun Palace was an actual part of the design process, functioning as a means to achieve its social ends.

Cybernetics in the Fun Palace and Yona Friedman's Flatwriter

Cybernetics (as well as game theory), as a discipline that could deal with unstable, indeterminate, and evolutionary systems, provided the methods to incorporate computational technologies in the Fun Palace. Mathews explains:

> Using cybernetics and the latest computer technologies, Price hoped to create an improvisational architecture which would be capable of learning, anticipating, and adapting to the constantly evolving program. An array of sensors and inputs would provide real-time feedback on use and occupancy to computers which would allocate and alter spaces and resources according to projected needs.[73]

Cybernetics pioneer Gordon Pask, who was in charge of the cybernetic systems in the Fun Palace, in his article "The Architectural Relevance of Cybernetics," described how the cybernetic paradigm of design could link the aim or intention of users, with programs about how the system could start to learn, adapt, and respond to these aims and intentions.[74] Specifically, the cybernetic system in the Fun Palace would collect and analyze information about the preferences and individual activities of visitors, computing overall tendencies and determining parameters for the way space would transform (moving walls and corridors). Thus, the Fun Palace would be able to "learn" behavioral patterns and plan future activities, anticipating phenomena, changing tendencies and possible uses.[75]

Two basic electronic systems would be built into the Fun Palace to achieve these aims: the first was a system of perforated computerized punch cards that registered information on the kind, as well as the size, location, quality and quantity of activities. This system would be able to monitor activity patterns, avoid programmatic conflicts and distribute resources for different activities, such as televisual images, telecommunications, electricity, air-conditioning, lighting levels and possible functions. The second system was what Roy Ascott, a member of the cybernetics committee, called the *Pillar of Information*, which would act as an electronic pavilion, able to store and represent a variety of information, modeled on Encyclopedia Britannica. This system would be able to register use patterns and past interactions, storing in its memory the most interesting cases of specific users. In this way, it would be able to create a non-hierarchical information map, inducing and expanding searching activity by users.[76]

The Fun Palace would react to patterns of use as they appeared, and provoke new ones by means of adaptive learning mechanisms. Pask had already designed and implemented similar mechanisms in his earlier work with computational, theatrical and learning machines, such as his *Musicolor machine* (which we will further discuss in Chapter 4). By contrast to first-order cybernetic systems that self-adjust towards a predetermined and fixed goal, Pask's cybernetic mechanisms were modeled on second-order cybernetics. This was an information discourse proper to the emerging post-industrial society and knowledge-based economy,[77] which Price wanted to anticipate. In the context of second-order cybernetics, observers are inseparable from the system being observed (the machine) and with which they interact, configuring and reconfiguring their goals dynamically. Observer and observed participate in open-ended reciprocal exchanges – what Pask called conversation, a cybernetic concept that runs through all of Pask's work, both applied and theoretical.[78]

A similar approach to human–machine interaction for the participatory formation of architecture, albeit with certain differences, was proposed by Yona Friedman. The Hungarian architect designed a conceptual machine, the *Flatwriter*, presented in the form of diagrams and explanatory notes, that would assist in configuring the so-called *Ville Spatiale*, his proposal for a user-determined

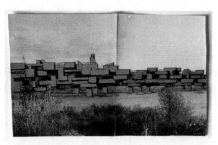

FIGURE 1.12 *Ville Spatiale*, Yona Friedman. Yona Friedman Archives, Paris © Courtesy Yona Friedman, all right reserved / ADAGP, Paris / OSDEETE, Athens 2015.

residential megastructure (Figure 1.12). The Flatwriter would not only receive input from users, but also respond to their choices to optimize the resulting outcome. More specifically, it would prompt potential future residents to choose the plan and location of their houses within the unchanging infrastructure of the *Ville Spatiale*, help optimize the resulting outcome (by evaluating these choices and assessing conflicts with other existing houses), and inform constructors and the other residents of the changes and their consequences.

The Flatwriter comprised two keyboards: one for configuring the plan of the apartment, as suggested by the user, and the other for choosing user habit frequencies (Figure 1.13). The first keyboard consisted of 53 keys that represented a number of configurations of a three-room apartment, room shapes, different locations of prefabricated equipment units within the rooms, and alternative orientations of the house. Each future resident could print his preferred apartment (along with a computed cost), after selecting, in four successive steps (using 8 to 9 keys), how he wanted the plan to be configured, the shape of each of the three rooms, the location of bathroom and kitchen equipment within these rooms, and finally the orientation of the house. According to Friedman, the system could offer some millions of possible plans. In the next stage, using the second keyboard, the user should input information about his or her domestic habits, by selecting a key according to their knowledge of how often they went into each of these rooms.

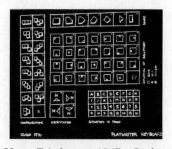

FIGURE 1.13 *Flatwriter*, Yona Friedman, 1967, Osaka project. Yona Friedman Archives, Paris. © Courtesy Yona Friedman, all right reserved / ADAGP, Paris / OSDEETE, Athens 2015.

In the next step, the Flatwriter would assess this user input and respond by warning the future resident about issues such as tiredness, comfort, isolation or communication. After final decision and selection, the system would present on a TV screen a grid of the project infrastructure, showing already occupied, as well as unoccupied areas, where the apartment could be inserted. The user would then specify a preferred location for the house, and the system would subsequently check whether this choice obstructed access, natural light and ventilation of the neighboring flats. In case of disturbance, the system would prompt the user to choose again. In the final stage the system would inform all residents about overall changes induced by the new entry, in terms of intensity of motion, noise or air pollution.[79]

As a feedback system, and despite interaction between user and machine, the Flatwriter was a first-order cybernetics model, because it was constantly trying to eliminate error; its aim was to minimize disturbances in the life of the city, and the potential conflict with neighboring houses, by returning warnings to the users, prompting them to modify their choices, in order to optimize the overall layout.

But, as already mentioned, the Fun Palace was a different story. Although the role of conversation in interactive environments will be further addressed in Chapter 4, it is for now sufficient to say that the Fun Palace was a sort of "performative" space, because, as Lobsinger explains, content, use and program were performed through moments of interaction between users and the machine.[80] Pask, in his text about the relevance of cybernetics to architecture, proposed this conceptualization of architecture as a partner that can converse with users:

> The high point of functionalism is the concept of a house as a "machine for living in." But the bias is towards a machine that acts as a tool serving the inhabitant. This notion will, I believe, be refined into the concept of an environment with which the inhabitant cooperates and in which he can externalize his mental processes.[81]

Yet, as Mathews argues, for Pask, the role of cybernetics in architecture was not just to create interactive servicing environments but also to control human behavior; every parameter would be quantified and mathematically modeled into information-data, registered in flowcharts (the relation with the environment, the interactive activities, the mechanical capacities and parameters, the learning machines, the controllable team activities, the communication and information systems, the air-conditioning systems and the artistic forms) (Figure 1.14). On the other hand, Andrew Pickering thinks that Mathews's critique on the tendency for control in the Fun Palace is misdirected, because the Fun Palace was part of the cybernetic practices and discourse in Britain in the 1940s and 1950s. In this environment, scientists constructed electromechanical machines that manifested conversational and adaptive behavior that staged, as Pickering

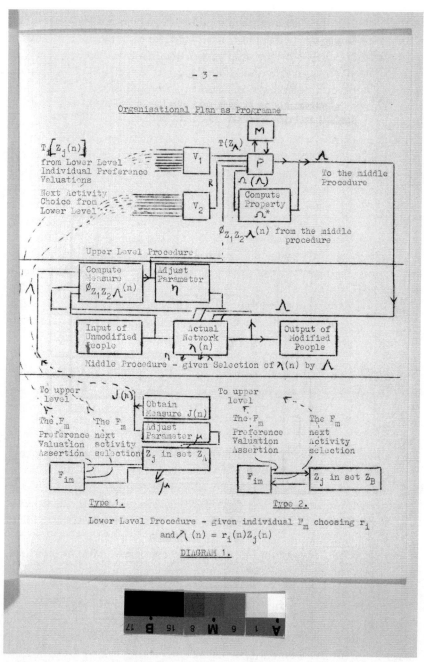

FIGURE 1.14 "Organisational Plan as Programme," from the minutes of the Fun Palace cybernetics committee meeting, 27 January 1965, Gordon Pask, author, Cedric Price, architect, 1965, reprographic copy on paper, 25.6 × 20.5 cm. Cedric Price Fonds, Collection Centre Canadien d'Architecture/Canadian Centre for Architecture, Montréal.

argues, a sort of non-modern ontology. In this perspective "the Fun Palace was just another staging of that ontology."[82]

Although the form of the Fun Palace, with its mechanical systems, revolving escalators, moveable hinged roof, cranes and trusses, could bring to mind a functionalist and determinist attitude to flexibility, the aforementioned uncertainty in terms of both program and form points to a different type of machine as a metaphor for architecture. Mathews argues that the Fun Palace can be considered to be what mathematician Alan Turing called a "universal machine," an entirely conceptual device that he described in a seminal paper in 1936. This device was an early anticipation of the modern electronic computer, which can simulate the function of several machines, such as the typewriter or the blackboard, depending on programming.[83] Thus, as Mathews explains, the Fun Palace can be thought of as a virtual and thus, flexible architecture, able to emulate different functions of buildings.[84]

The open-endedness and indeterminacy implied here, in contemporary terms, point to the workings of the philosophical concept of the *virtual*, addressed by Gilles Deleuze, following Bergson, to describe a not-yet-effectuated concrete reality.[85] Thus, although the Fun Palace was an idea that remained on paper,[86] it can inspire us to think of architecture as a functionally open-ended adaptive environment, beyond the determinist limitations of mechanical flexibility. To study this further, we will later look into recent examples of such mechanical structures, in which digital technology sometimes is used to enhance their kinetic and responsive capacities.

Takis Zenetos's Electronic Urbanism

The editors of Archigram 5 (1964), in a page titled "Within the Big Structure," juxtaposed a picture of Archigram's *Plug-in City*, a model of Constant Nieuwenhuys's *New Babylon*,[87] and a single perspective sketch of Greek architect Takis Zenetos's floating megastructure, which he termed "Electronic Urbanism" in 1969. Zenetos was a member of the Association Internationale de Cybernetique, and, as a visionary architect, he was concerned with the role of the emerging communication and information technologies in the design of cities and architecture within the post-war socio-economic environment. Like Price, he believed in the potential of science, technology and cybernetic thinking for social change, although he was not interested in expendability, throwaway architecture and pop culture, like the Archigram group. But like the latter, he rejected the monumentality of structures that, as he thought, limited human freedom; instead, he favored fluid, "open" and decentralizing infrastructures and technological equipment (such as prefabricated plastic capsule homes, or mass-produced building components selected and assembled by users) that could reflect, facilitate and anticipate the constantly changing human needs.[88]

Zenetos's *Electronic Urbanism* project, first conceived, as he said, in 1962, was published and exhibited in several shows and periodicals in Greece,[89] culminating in his 1969 trilingual monograph titled *Urbanisme électronique: structures parallèles / City Planning and Electronics: Parallel Structures /* Ηλεκτρονική πολεοδομία: παράλληλες κατασκευές (Athens: Architecture in Greece Press, 1969).[90] By contrast to Archigram's and Price's megastructural schemes that interfered with the existing city, Zenetos's floating infrastructure was freed from the constraints of the existing city,[91] a modernist de-territorialization of architecture, such as Yona Friedman's Spatial City and Constant's New Babylon. At the same time, however, he opposed the coexistence of industry and housing in Friedman's Spatial City, and the utopian future of a ludic society proposed by Constant. Moreover, unlike Archigram, Friedman and Constant, Zenetos explored the technical and quantitative details of his project (such as calculations of wind loads and earthquakes).

Eleni Kalafati and Dimitris Papalexopoulos discern three stages of development of the project from 1962 to 1974. The first was concerned mostly with prefabrication and flexibility within a web-like structure suspended all around the planet (Figures 1.15 and 1.16).[92] In the second stage, Zenetos addressed the problem of growth of cities due to the expanding tertiary economy that, as he believed, ruptured the social functions of the city, and caused the relocation of housing outside the city, thus increasing the distance between residence and work. He had the view that the emerging problems of cities were driven by the expanding service sector, the supporting transportation infrastructure and the suffocating overpopulation.[93] As a potential solution he proposed the elimination of the need for transportation to the places of tertiary activity, and the use of telecommunications to automate services and work from home (tele-work) (Figure 1.17). In the third stage information technology was much more prevalent in his drawings which depicted the prospect of an immaterial future – a non-architecture (Figure 1.18).[94]

Zenetos's drawings of the adaptable equipment embedded on the megastructure, recall similar ideas in the work of Archigram, Banham and Price. His *Spinal Body Carrier* would be a sort of adjustable prosthetic second body, like Archigram's *Cushicle and Suitaloon*, equipped with audio-visual interfaces and telecommunication systems, to facilitate creative and work-related activities, distant meetings and tele-cooperation services for its occupant. It would also facilitate recreation, or "play," as Zenetos called it, for the members of the future "leisure society."[95] Its hemispherical double-skin envelope would have an adjustable degree of opacity enabled by the flow of a controllable amount of a transparency regulating liquid in the void between the two skins (Figures 1.19 and 1.20).[96] This is very similar to the environmental barriers of warm or cold air that Price and Banham proposed for the Fun Palace and the Environment Bubble respectively. And just like Archigram, Zenetos's thinking gradually moved from suspended megastructures and adjustable equipment to the vision of an immaterial future, with holograms and techniques to control wave bands and fields, providing the desired environmental conditions at an instant.[97] Zenetos even suggested the use of display-walls, namely full-color TV screens with

Architecture, Adaptation and the Cybernetic Paradigm 35

FIGURE 1.15 The structure provides sufficient transparency, from "Architecture in Greece" 3 (1969), Takis Zenetos. © Reproduction permission granted by the Hellenic Institute of Architecture.

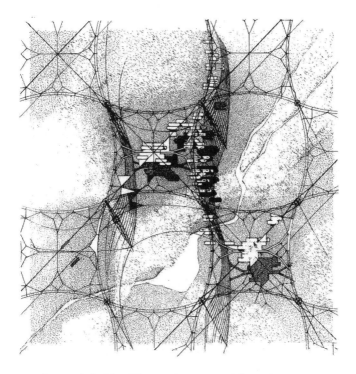

FIGURE 1.16 Suspended City. The environment. The urban structure co-exists with nature, from "Architecture in Greece" 4 (1970), Takis Zenetos. © Reproduction permission granted by the Hellenic Institute of Architecture.

36 Architecture, Adaptation and the Cybernetic Paradigm

FIGURE 1.17 Vegetation extending over two levels, from "Architecture in Greece" 8 (1974), Takis Zenetos. © Reproduction permission granted by the Hellenic Institute of Architecture.

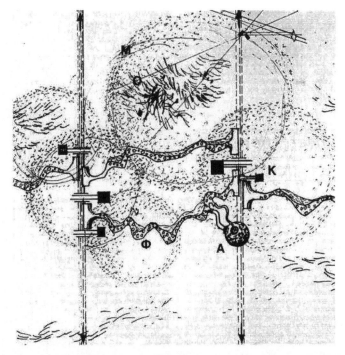

FIGURE 1.18 Immaterial Architecture (distant future), from "Architecture in Greece" 8 (1974), Takis Zenetos. © Reproduction permission granted by the Hellenic Institute of Architecture.

Architecture, Adaptation and the Cybernetic Paradigm **37**

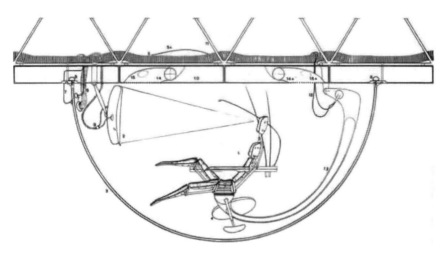

FIGURE 1.19 Tele-processing cell, from "Architecture in Greece" 8 (1974), Takis Zenetos. © Reproduction permission granted by the Hellenic Institute of Architecture.

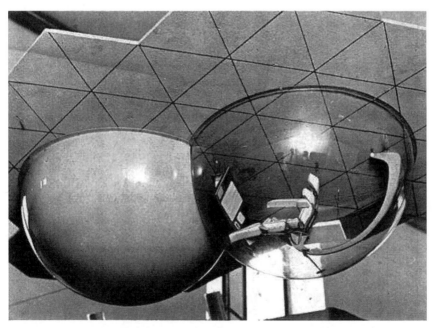

FIGURE 1.20 Twin-cell envelope, of variable transparency, for two tele-activities assemblies, from "Architecture in Greece" 8 (1974), Takis Zenetos. © Reproduction permission granted by the Hellenic Institute of Architecture.

three-dimensional high precision output picture, that would connect people in different physical locations and encourage active participation of viewers in an "infinite number of tele-activities."[98] Therefore, Zenetos's *Electronic Urbanism* anticipated current applications of digital technology in the built environment, a vision which, as we shall see in Chapter 2, is embedded in the research projects of so-called ambient intelligence.

Kinetic Structures and Adaptive Systems

Kinetic structures use moveable mechanisms, and sometimes actuator devices and computational technologies, to effect changes in the form and configuration of buildings and spaces. Although they seem to be adaptable structures, as we shall see, they implement a determinist, "hard" type of flexibility. Michael Fox, founder of the Kinetic Design Group at MIT (1998–2000), and Bryant Yeh have defined kinetic structures as "… buildings or building components, with variable location or mobility and/or variable geometry or movement."[99] Such structures are called by different names, such as mobile, portable, transformable, convertible and deployable, depending on whether movement and change is a feature of the whole structure or parts of it. Of course, although structures with lightweight, deployable, and transformable components have existed since antiquity, they were more widely developed in the second half of the 20th century as a response to the rapid changes in western societies, and to the economical, practical or ecological problems that the construction industry had to cope with.[100] More recent examples, however, are part of a wider experimental research direction; these include the kinetic projects of Hoberman associates and Buro Happold, the computationally-driven structures of the Hyperbody research group at TUDelft and some projects by Michael Fox and his former Kinetic Design Group at MIT.

In most of these cases, kinetic structures are physical constructions consisting of moveable interconnected parts that can rearrange their relative positions, according to demand, either manually or through feedback control systems. The result is an overall change of the physical configuration of the structures, which is determined by the set relations of their internal components and their inbuilt kinetic mechanisms. The latter may range from mechanisms of deployment, folding and extension, to rolling, sliding and nesting techniques, and from scissor-type mechanisms and inflatables, to tensile systems embedded with electromagnetic, pneumatic or hydraulic actuators (Figure 1.21).

Although kinetic systems are part of a research project with many more aims than just flexibility, our investigation focuses on whether and how kinetic structures can adapt and respond to change.[101] In an early book on kinetic architecture, William Zuk and Roger Clark argued that the aim of kinetic solutions is to make spaces adapt to change; this means that architects should conceptualize architecture as "a three-dimensional form-response

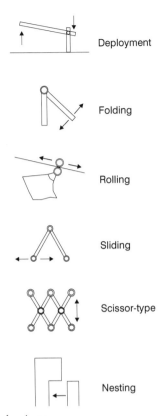

FIGURE 1.21 Kinetic mechanisms

to a set of pressures,"[102] namely functional, social or environmental factors that may influence its form: "We must evolve an architecture which will adapt to continuous and accelerating change – a kinetic architecture... Basic to the philosophy of kinetic architecture is the importance of being able to accommodate the problem of change."[103] Similarly, Richard Rogers, writing about future developments in design, has explained:

> Present-day concerns for static objects will be replaced by concern for relationships. Shelters will no longer be static objects but dynamic objects sheltering and enhancing human events. Accommodation will be responsive, ever changing and ever adjusting.[104]

Later, Michael Fox and Bryant Yeh argued that the central motive of kinetic solutions in architecture (either in residential, entertainment, educational or commercial buildings), and the application of computational technologies in architecture since the 1990s, is to make spaces able to respond to constantly changing needs of contemporary life, and to optimize their performance, providing energy efficiency, functional adaptation, security and mobility.[105]

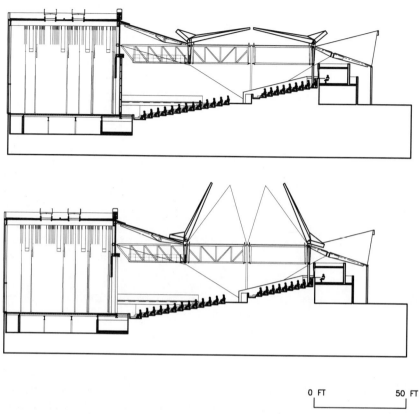

FIGURE 1.22 Section of Bengt Sjostrom Starlight Theatre. © Courtesy Studio Gang Architects.

In reality, kinetic techniques are mostly employed to construct moveable building components, such as transformable skins and facades, sliding roofs and entrances. Some examples include the kinetic roof of the entrance lobby of the Milwaukee Art Museum (Quadracci Pavilion) by Santiago Calatrava, the opening roof of the Bengt Sjostrom Starlight Theater by Studio Gang Architects (Figures 1.22 to 1.24), and the *Iris Dome* or the *Hoberman Arch* by Hoberman Associates, which was inspired by Emilio Pérez Piñero's early deployable structures. Other more experimental approaches however have been employed to deal with interior domestic environments and how they interface with human users. For instance, a few years ago, the Hyperbody research group built a number of experimental 1:1 spatial prototypes, the *Muscle projects*, as part of their applied research in the BSc/MSc courses. These included the *Muscle Re-Configured*, the *Muscle Room*, the *Muscle Body* and the *Muscle Space*, which followed ONL's exhibit, the *Muscle NSA*, at the Non-Standard Architecture Exhibition at the Centre Pompidou in 2003.[106]

The *Muscle Reconfigured* (2004), essentially a reconfigured version of the *Muscle NSA*, was a transformable prototype spatial structure, a three-dimensional strip

FIGURE 1.23 Interior image of Bengt Sjostrom Starlight Theatre. © Gregory Murphey.

FIGURE 1.24 Interior image of Bengt Sjostrom Starlight Theatre. © Gregory Murphey.

made of Hylite® panel sheets,[107] able to bend by means of the compression power produced by a pair of Festo Muscle pneumatic actuators[108] attached on each of them. Actuation is effected by processed data input from proximity and touch sensors, making individual panels as well as the overall shape of the strip deform.[109] The *Muscle Room* (2005), on the other hand, was more elaborate in terms of functional capacities. Its designers envisioned it as a hotel room able to convert its walls and floor into different furniture items according to need. It was made up of a wooden frame, holding a secondary kinetic system, with rotating joints and angle lock mechanisms, driven by pneumatic actuators and bicycle cables.[110] The *Muscle Body* (2005) consisted of a stretchable Lycra skin, enveloped and stiffened by a spiraling PVC tube, bent in three dimensions, on which 26 industrial Festo muscles were attached. Imbedded pressure and proximity sensors provided the data input for the synchronized actuation of the Festo muscles, leading to a variation of the translucency of the skin (according to the degree of stretching), and the play of internal light.[111] The same PVC tubes were used as a structural flexible grid system in the *Muscle Space* (2007) project. The structure, which was conceived as a passageway for visitors, comprised a number of pneumatic Festo muscles able to induce movement to the entire object. When visitors passed through the space, pressure sensors laid on the floor registered the location of their footsteps to provide data input for the actuators.[112]

The apparent rigidity of these kinetic structures is striking. The position and number of actuators, the physical limits of their contraction, the extent to which the metal sheets, skin and joints can bend, stretch or rotate respectively, as well as the very internal structure and articulation of the kinetic system, determine and constrain the transformational capacities of the projects, and hence their flexibility. Indeed, kinetic structures involve a determinist approach to flexibility, because the determinable degree of freedom of movement that they are able to produce means that the range of possible functional changes that may occur is predictable. This is because the structures are limited by what Zuk and Clark call closed-system kinetics – the form can only "respond to a range of functional changes possible within the initial envelop limitations."[113] Therefore, although flexibility depends on the multiple transformational states of the structure, it is confined within a narrow range of alternatives, predefined by design.[114]

From Mechanical Machines to Adaptive Systems

The work of the architectural avant-garde was modeled on the notion of open-ended and adaptive cybernetic machines. But the aforementioned kinetic structures are modeled on the logic of a mechanical machine, such as that addressed by Canguilhem in the lecture we mentioned at the beginning of the chapter. Mechanical machines, as he explains, follow a traditional theory of kinematics, such as that in Franz Reuleaux's *Theoretische Kinematik: Grundzüge einer Theorie des Maschinwesen* (Braunschweig: Vieweg, 1875). Canguilhem explains that in these theories, mechanical movement holds up to certain norms, measures or estimates. Mechanical systems are made up of a set of moveable components that work together with calculated precision and periodically return to their predetermined set relations. They are not capable of self-adjustment, unpredictable behavior, learning, adaptation or "virtual" functions (open-endedness), which are functional characteristics of living organisms.[115] It is this discrepancy between the mechanical and adaptive machine, as models for architecture – between an actualized reality and a virtual future – that seems to run through the vision and drawings of the architectural avant-garde discussed in the previous sections.

What we want to investigate then is whether this vision is (or can be) possible today. The pervasive presence of digital technology in our everyday environments the last 25 years, has raised an interest in its potential integration in the built environment and the subsequent augmentation of architecture with adaptive properties. Architecture historian Antonino Saggio, for instance, has argued that information technology provokes a potential paradigm shift in architecture and its historical role since modernism: "We no longer speak of *Existenz Minimum*, for an architecture that satisfies needs, but if anything of *Existenz Maximum* for an architecture that expands possibilities and desires."[116]

This differentiation between need and desire problematizes the concept of need and its functionalist connotations, which is reflected in several writings of urban sociologists and architectural theorists since the 1960s. Lefebvre discusses the concept of need on the grounds of the Marxist critique of everyday life, coinciding with the view of other Marxist theorists, such as Herbert Marcuse and Jean Baudrillard, who also criticized the concept of need within the practices of the capitalist economy and the relations of production and consumption. Lefebvre's critique develops the idea of social needs, reflecting the cultural and social development of individuals and groups, which also follows some of the lines of 1960s architectural theory and the work of urban sociologist Paul-Henry Chombart de Lauwe.[117] As Łukasz Stanek explains, these views re-introduced cultural and political choices in architecture, and problematized the concept of need as defined by merely scientific criteria in functionalist modernism.[118]

Of course, Archigram's consumerism-inspired projects, designed to adapt to changing human desires, can hardly be considered to follow this polemical attitude,[119] unlike maybe other futuristic projects of the same period such as Constant's New Babylon. And this is what we will see in the next chapter, where we will discuss intelligent environments that seem to actualize Archigram's ideas. But still, the concept of desire seems to be linked to the potential of information technology to offer multiple choices in people's life, especially if we consider the way large IT corporations manage and sell certain lifestyles, through the way they promote and advertise gadgets. In the context of architecture, for Saggio, as well as for Dimitris Papalexopoulos,[120] the information technology paradigm may invite a new approach to the design of architecture, concerned with its capacity to respond to the subjectivity of human desires:

> Attempts are being made to understand how to make the computer interact with us humans and the environment by using various types of sensors: mechanical and quantitative ones that measure air, light or temperature; more complex ones able to interpret facial expressions or a tone of voice; and others even more sophisticated that manage to formulate hypotheses about what we might emotionally "feel." For a small, but essential, group of architects, this frontier is also a fundamental attempt to make architecture *change interactively with changing situations and desires*.[121]

This vision is the subject of our next chapter, which will look at the fields of Artificial and Ambient Intelligence, to discuss the attempts that engineers and architects have made towards the potential fusion between information technology and architectural space.

Notes

1 See George Canguilhem, "Machine et Organisme," in *Connaisance de la Vie* (Paris: Hachette, 1952). Reprinted in George Canguilhem, "Machine and Organism," in *Incorporations*, eds. Jonathan Crary, and Sanford Kwinter (New York: Zone Books, 1992), 44–69.
2 See André Leroi-Gourhan, *Evolution et techniques II: Milieu et techniques* (Paris: Albin Michel, 1945). The theory of organic extension was first articulated by the German writer Ernst Kapp in his *Grundlinien einer Philosophie der Technik* (Braunschweig: Westermann, 1877), although its philosophical foundations go back to Eduard von

Hartmann's 1869 book *The Philosophy of the Unconscious,* and even further back to Arthur Schopenhauer. French and mainly German philosophers disseminated the philosophical path of this conceptualization at the end of the 19th and the beginning of the 20th century; Alfred Espinas, who borrowed the theory of organic extension from Ernst Kapp to explain how the first tools were constructed, and Alard Du Bois-Reymond and Oswald Spengler, who grounded their theories on technological invention on Darwinian ideas about variation and natural selection. See Canguilhem, "Machine and Organism," 61.

3 Canguilhem, "Machine and Organism," 44–69. Canguilhem made this remark in the last footnote of his essay, where he mentioned bionics, the study of biological structures and systems as models or analogues for technological ones, citing Jean Dufrenoy's 1962 article "Systèmes biologiques servant de modèles pour la technologie." This organic understanding of machines is a thought that Lewis Mumford had already discussed in a text titled "Toward an Organic Ideology" in his book *Techniques and civilization* (1934), in which he put forth a proposal for a new machine ethic attached to a more responsive form of technology, and influenced by Patrick Geddes' notion of biotechnics. Contrary to the organism-as-machine conceptualization, he suggested that machines are rather "lame counterfeits of living organisms," to undermine the authority of the purely mechanical. He added: "Our finest airplanes are crude uncertain approximations compared with a flying duck: our best electric lamps cannot compare in efficiency with the light of the firefly: our most complicated automatic telephone exchange is a childish contraption compared with the nervous system of the human body." See Lewis Mumford, "Technical Syncretism and Toward an Organic Ideology," in *Rethinking Technology: A Reader in Architectural Theory,* eds. William Braham and Jonathan Hale (London/New York: Routledge, 2007), 53–57.

4 Norbert Wiener, *Cybernetics or Control and Communication in the Animal and the Machine* (Cambridge, MA: The Technology Press, 1948).

5 Norbert Wiener, *The Human Use of Human Beings: Cybernetics and Society* (London: Free Association Books, 1989).

6 John Johnston, *The Allure of Machinic Life: Cybernetics, Artificial Life, and the New AI* (Cambridge, MA/London: MIT Press, 2008), 29–31.

7 Wiener, *The Human Use of Human Beings,* 26–33.

8 Eastman, "Adaptive-Conditional Architecture," 51–57.

9 Negroponte, *Soft Architecture Machines,* 144.

10 Michael Fox, "Beyond Kinetic," in *Gamesetandmatch Conference Proceedings* (Publikatiebureau Bouwkunde, Faculty of Architecture DUT, 2004), http://www.bk.tudelft.nl/fileadmin/Faculteit/BK/Over_de_faculteit/Afdelingen/Hyperbody/Game_Set_and_Match/GameSetandMatch_1/doc/gsm_I.pdf.

11 Negroponte, *Soft Architecture Machines,* 47–48.

12 See Heinz von Foerster, *The Cybernetics of Cybernetics* (Minneapolis: FutureSystems Inc, 1995 [1975]); Heinz von Foerster, "On Self-Organising Systems and Their Environment," in *Observing Systems* (Salinas CA: Intersystems Pubilcations, 1981 [1960]).

13 For instance, architects like Rem Koolhaas used autopoietic theory as a metaphorical model in urban design, because urban systems are complex systems that can adapt, reorganize themselves and grow, consisting of multiple layers of disorder and chaos. See Rem Koolhaas et al. (eds.) *Mutations* (Barcelona: Actar, 2000). On this, Christine Boyer noted that it is not clear whether the authors recognized that they borrowed the analysis of autopoietic systems from Maturana and Varela's theory to develop their own interpretative model of cities. But the way autopoietic theory was articulated, allowed its generalized use, and the possibility to create analogies between different systems, either biological urban or social. See Christine M. Boyer, "The Body in the City: A Discourse on Cyberscience," in *The Body in Architecture,* ed. Deborah Hauptmann (Rotterdam: 010 Publishers, 2006), 42. Also see Francisco Varela, *Principles of Biological Autonomy* (New York: Elsevier North Holland, 1979), 565. In this book, Varela pointed

out that both society and culture are, like an organism, systems in which a higher order system can emerge from the interactions between political, economic, cultural, communicative, legal, scientific and religious systems.

14 For instance see Stephen Gage, "Intelligent Interactive Architecture," *Architectural Design* 68, 11–12 (1998): 84; Gillian Hunt, "Architecture in the Cybernetic Age," *Architectural Design* 68, 11–12 (1998): 53–55; Ted Krueger, "Autonomous Architecture," *Digital Creativity* 9, 1 (1998): 43–47; Ted Krueger, "Like a second skin, living machines," *Architectural Design* 66, 9–10 (1996): 29–32; Ted Krueger, "Editorial," *Convergence: The International Journal of Research into New Media and Technologies* 7, 2 (2001); Ranulph Glanville, "An Intelligent Architecture," *Convergence: The International Journal of Research into New Media and Technologies* 7, 2 (2001): 15–16.

15 Adrian Forty, *Words and Buildings: A Vocabulary of Modern Architecture* (New York: Thames & Hudson, 2004), 144.

16 The Schröder Huis (1924) is an exemplary case of a flexible house. The architect Gerrit Rietveld designed a flexible first floor plan, able to respond to the specific and predetermined requirements and daily cycles of the client, Mrs Truus Schröder. The first floor plan could completely open up during the day, by sliding its hinged panels back towards the external walls in minutes; during the night the panels could be pulled back to their unfolded stated to divide the open plan into two separate bedrooms and one living/dining room, sharing a common central hall. Thus, despite an implied intention to create domestic space that would constantly change according to the needs of its inhabitants at different times of the day, the floor plan had only two states: open plan and divided. According to Catherine Croft, the overall function of the plan was authoritative, because it exercised control over the family life, which is not aligned to the idea of freedom usually attributed to the concept of flexibility. See Catherine Croft, "Movement and Myth: The Schröder House and Transformable Living," *Architectural Design: The Transformable House* 70, 4 (2000): 12–13. See also Tatjana Schneider and Jeremy Till, *Flexible Housing* (Oxford: Architectural Press, 2007), 57, 152.

17 For this housing project (1934), Johannes Van den Broek designed a flexible efficient organization in plan that could be altered on a daily basis (with moveable walls and furniture as well as overprovision of doors), without the sacrifice of comfort. The architect based his work on systematic surveys of use-cycles of residential space that he conducted with his collaborator Heinrich Leppla, and the plans that evolved from these studies. During the day, the central area could be opened up to a single space, and during the night, it could be divided into two separate rooms by sliding panels and fold-down beds, while the study room could also be turned into a bedroom. Routes and activities in the house were not hindered by these alterations, because the plan was overprovided with doors to serve each different layout. But although the plan could take on 24 different ways of organizing activities, its flexibility was still predetermined by design. It is questionable, therefore, whether this project can be placed within the "soft" category of flexibility, as Schneider and Till seem to think (this is at least how they classify the project in its webpage at http://www.afewthoughts. co.uk/flexiblehousing/house.php?house=18&number=&total=&action=&data =&order=&dir=&message=&messagead=). Also see Schneider and Till, *Flexible Housing*, 65.

18 Le Corbusier designed the Maisons Loucheur (1928/29) as a response to a French governmental housing program for a total of 260,000 dwellings. Building on the idea of daily changing usages that he had already explored in other buildings, Le Corbusier conceived them as prefabricated housing units for a family of up to four children that would be able to expand the 46m^2 plan to 71m^2 during the 24-hour daily cycle, by means of embedded moveable walls and fold-down furniture, and thus to maximize the number of functions in the same space. See Schneider and Till, *Flexible Housing*, 61.

19 Schneider and Till, *Flexible Housing*, 7.

20 Forty, *Words and Buildings*, 143.
21 In the context of Taylorism, the dominant rationalist discourse and perhaps the most emblematic social technology of the early 20th century, the factory worker was pervasively implied as a machine, a single functional and predictable unit of movement. The man-as-machine metaphor justified the techniques used for human engineering, such as the chronocyclegraphs described in Frank and Lillian Gilbreth's *Time and Motion* studies. These techniques were used to record, measure and regulate the workers' micromotions, in order to optimize the rational organization of industrial production. These practices, in the architectural discourse of the 1930s, led to the regulation and functional optimization of the building – the house-as-machine. Examples can be found in the post-war state-funded housing programs, which involved scientific management techniques, guided by low-cost, functional optimization and efficiency. For instance, in Schutte-Lihotzky's "Frankfurt Kitchen" (1927), and Alexander Klein's "Functional House for Frictionless Living" (1928), normalizing techniques like the plan, regulated and optimized human standardized functions and movements. Behaviorism (the most prevalent branch of psychology of that period) and social engineering, led this mechanistic view that society and individuals were subject to manipulation and control. As Jeremy Till explains, these practices fitted the more general pattern of the will to order which, in Zygmunt Bauman's sociological theory, is considered to be a central feature of modernity. See: Jeremy Till, *Architecture Depends* (Cambridge, MA/London: MIT Press, 2009), 33–34. For an account of the processes of Taylorism in industry, time and motion studies, and their wider cultural implications, see Siegfried Giedion, "Industrialization as a Fundamental Event," in *Rethinking Technology: A Reader in Architectural Theory*, eds. William Braham and Jonathan Hale (London/New York: Routledge, 2007), 75–105.
22 See for instance Alan Colquhoun, "Plateau Beaubourg," in *Essays in Architectural Criticism: Modern Architecture and Historical Change* (Cambridge, MA/London: MIT Press, 1981), 110–119.
23 Forty, *Words and Buildings*, 143.
24 Schneider and Till, *Flexible Housing*, 7.
25 Schneider and Till discuss this distinction citing Steven Groak's book *The Idea of Building* on this matter. See Schneider and Till, *Flexible Housing*, 5.
26 See Jonathan Hill, *Actions of Architecture: Architects and Creative Users* (London: Routledge, 2003), 37–43.
27 Forty, *Words and Buildings*, 144.
28 See Herman Hertzberger, *Lessons for Students in Architecture* (Rotterdam: 010 Publishers, 1991), 147. A similar concept is Bernard Leupen's *generic space*, which points to the changeable indeterminate area in buildings that can be either alterable, when it contains a layer with physically changeable parts, extendable, or polyvalent. In his book *Frame and Generic space* (2006), following the succession of studies in changeability conducted by the Foundation for Architectural Research (SAR) and others, Leupen reviewed many 20th-century buildings, analyzing their potential for flexibility and changeability, by emphasizing the permanent layer of their organization. To do this, he proposed an analysis of buildings by decomposing them into a system of five layers – structure, skin, scenery, services and access. Any of these layers can acquire the role of the permanent layer, the "frame," either alone or in combination with others. The "frame" layer defines the "generic space," i.e. the unspecified space in buildings, which is subject to adaptability and change. See Bernard Leupen, *Frame and Generic Space* (Rotterdam: 010 Publishers, 2006).
29 Forty, *Words and Buildings*, 148.
30 For, probably, the most valid explanation of Lefebvre's spatial triad see Christian Schmid, "Henri Lefebvre's Theory of the Production of Space: Towards a three-dimensional dialectic," in *Space, Difference, Everyday Life: Reading Henri Lefebvre*, eds. Kanishka Goonewardena et al. (London/New York: Routledge, 2008) 27–45.

31 Henri Lefebvre, *The Production of Space*, trans. Donald Nicholson-Smith (Oxford: Blackwell, 1991), 369.
32 Ibid., 362.
33 For instance see the work of Hertzberger in the 1970s and Jonathan Hill's edited book *Occupying Architecture* (London: Routledge, 1998) and *Actions of Architecture: Architects and Creative Users*. Also see Daniel Maudlin and Marcel Vellinga, eds., *Consuming Architecture: On the Occupation, Appropriation and Interpretation of Buildings* (London: Routledge, 2014). Jonathan Hill proposes the idea of active and creative users, when he discusses the significance of the analogy between writer-text-reader and architect-building-user (although he makes clear that texts cannot be compared to buildings). Paraphrasing the title of Roland Barthes's text *The Death of the Author* (1977), he proposes, not the death of the architect, but a new architect who takes into account the fact that users are creative agents, and that they construct a new building through use, in the same way that readers, as Barthes argued, make a new text through reading. See Hill, *Actions of Architecture*, 71–72.
34 Forty, *Words and Buildings*, 312–315.
35 Ibid., 148.
36 In a less political manner than Lefebvre, and more interested in the power of narratives, stories and myths, Michel de Certeau, in his book the *Practice of Everyday Life* (1984), examined ordinary practices, "tactics," as he called them, such as speaking, walking, cooking, or dwelling. People, in everyday situations, re-appropriate the imposed institutional order of culture, manipulating and conforming to the grid of "discipline," only to evade it, transforming and adapting it to their own interests. In this framework, the house, which can also be seen as a geography of imposed order, acquires stories of everyday practice, "tactics" that describe possibilities of use and everyday activity in its rooms, that is, "treatments of space" that defy and resist pre-established structures. See Michel de Certeau, *The Practice of Everyday Life* (Berkeley: University of California Press), 121–122.
37 As Stuart Brand explains, changing cultural currents, changing real-estate value, and changing usage, are the main factors that affect the way buildings transform in time. This characterizes most buildings, and even when they are not allowed to (due to legal limitations), although with different rates of change or cost. See Stewart Brand, *How Buildings Learn: What happens after they're built* (New York: Penguin, 1995), 2–10.
38 Charlie Gere, *Digital Culture* (London: Reaktion Books), 80–81.
39 Such as his book *Theory and Design in the First Machine Age* (London: Architectural Press, 1960), and his "stocktaking" series of articles. In *Theory and Design in the First Machine Age,* Banham assessed the association between the machine, technology and architecture in the modern movement. He deconstructed the stereotypical functionalist dogma of modernism, arguing that modern architects were mainly interested in aesthetics, style and formalism, and distanced themselves from Futurists and their enthusiastic embrace of technology. Instead they kept close ties with the academic tradition of Guadet and Blanc, and 19th-century rationalist theories. Looking at the most significant theoretical developments of the beginning of 20th century, which led to the "International style" of the 1930s and 1940s, and the role of the machine in the formulation of these developments, Banham suggested that Modern architecture "…produced a Machine Age architecture only in the sense that its monuments were built in a Machine Age, and expressed an attitude to machinery – in the sense that one might stand on French soil and discuss French politics, and still be speaking English." See Banham, *Theory and Design in the First Machine Age*, 329. But in the last chapter of his book, Banham wrote that the failure of modern architects to catch up with the constant evolution of technology, led to their decline since 1927, when Richard Buckminster Fuller constructed his revolutionary light-metal and mobile *Dymaxion House*. He observed that Fuller's Dymaxion House, with its lightness, expendability, industrial materials and mechanical services, was very close to the Futurist vision – for which Banham declared his particular liking –

and it was the only early example that came to terms with industrialized processes. See ibid., 327.

40 Although the Archigram group members used the term, it was John Weeks (of Llewelyn-Davies and Weeks architects) that used it first in an architectural context, and in a series of lectures he gave from 1962 to 1964. Weeks's case study was his Northwick Park Hospital (1961–1974), where he applied a strategy of open-ended hospital wings. See Jonathan Hughes, "The Indeterminate Building," in *Non-Plan: Essays on Freedom Participation and Change in Modern Architecture and Urbanism*, eds. Jonathan Hughes and Simon Sadler (Oxford: Architectural Press, 2000), 90–103. Also see John Weeks, "Indeterminate Architecture," *Transactions of the Bartlett Society* 2, (1962–64): 83–106; Richard Llewelyn-Davies, "Endless Architecture," *Architectural Association Journal*, (July 1951): 106–112. For Archigram's definitions of indeterminacy see Peter Cook et al., "Indeterminacy," *Archigram Eight: Popular Pak Issue* (1968), http://archigram.westminster.ac.uk/magazine.php?id=103&src=mg.

41 Peter Cook et al., "Exchange and Response," *Archigram Eight: Popular Pak Issue* (1968): http://archigram.westminster.ac.uk/magazine.php?id=103&src=mg. Reprinted in Peter Cook, ed., *Archigram* (New York: Princeton Architectural Press, 1999), 80.

42 These proposals included Kisho Kurokawa's 1961 *Helicoids* project and Arata Isozaki's 1960 *Joint Core System*. But by contrast to Archigram's emphasis on the liberating potential of plug-in architecture, the Metabolists' megastructural frameworks were tied to processes of bureaucratic planning and socio-political control. See Thomas Leslie, "Just What Is It That Makes Capsule Homes So Different, So Appealing? Domesticity and the Technological Sublime, 1945 to 1975," *Space and Culture* 9, 2 (May 2006): 180–194.

43 Alan Colquhoun, *Modern Architecture* (Oxford/New York: Oxford University Press, 2002), 225–226.

44 Peter Cook et al. "Emancipation," *Archigram Eight: Popular Pak Issue* (1968), http://archigram.westminster.ac.uk/magazine.php?id=103&src=mg. Reprinted in Cook, *Archigram*, 78.

45 Warren Chalk, "Living, 1990: Archigram Group," *Architectural Design* 69, 1–2 (January–February 1999): iv–v. Originally published in the March 1967 issue of *Architectural Design*.

46 Chalk, "Living, 1990," iv.

47 Peter Cook et al., "Metamorphosis," *Archigram Eight: Popular Pak Issue* (1968), http://archigram.westminster.ac.uk/magazine.php?id=103&src=mg. Also see Hadas Steiner, *Beyond Archigram: The Structure of Circulation* (London/New York: Routledge, 2009), 166.

48 Marshall McLuhan, *Understanding Media: the Extensions of Man* (Corte Madera, CA: Gingko Press, 2003 [1964]).

49 Peter Cook et al., "A very straight description," *Archigram Seven: Beyond Architecture* (1966), http://archigram.westminster.ac.uk/magazine.php?id=102&src=mg.

50 Peter Cook et al., "The Nomad," *Archigram Eight: Popular Pak Issue* (1968), http://archigram.westminster.ac.uk/magazine.php?id=103&src=mg.

51 Peter Cook et al., "Hard Soft," *Archigram Eight: Popular Pak Issue* (1968), http://archigram.westminster.ac.uk/magazine.php?id=103&src=mg. Also see Steiner, *Beyond Archigram*, 166–167.

52 See Peter Cook et al., "Room of 1000 Delights," and Ron Herron, "Holographic scene setter," *Archigram Nine: Fruitiest Yet* (1970), http://archigram.westminster.ac.uk/magazine.php?id=104&src=mg.

53 Steiner, *Beyond Archigram*, 33.

54 Hadas Steiner, "Off the Map," in *Non-Plan: Essays on Freedom Participation and Change in Modern Architecture and Urbanism*, eds. Jonathan Hughes and Simon Sadler (Oxford: Oxford Architectural Press, 2010), 135–136.

55 See Sadler, "Open Ends," in *Non-Plan*, 152. Also see Socrates Yiannoudes, "The Archigram Vision in the Context of Intelligent Environments and its Current

Potential," in *The 7th International Conference on Intelligent Environments (IE)*, Nottingham, 25–28 July 2011 (London: IEEE publications, 2011), 107–113.
56 Steiner, *Beyond Archigram*, 239.
57 See Simon Sadler, *Archigram: Architecture without Architecture* (Cambridge, MA/London: MIT Press, 2005), 197.
58 Banham, "A Home Is not a House," in *Rethinking Technology*, 117.
59 Banham borrowed the concept and term from Buckminster Fuller. It consists of TV screen, stereo speakers, solar power collector, refrigerator unit, electric cooker, audio player and other environmental gadgets.
60 Artist François Dallegret drew all these projects. See Reyner Banham, "A Home Is not a House," *Art in America* 2 (1965): 109–118.
61 Reyner Banham, "1960 1: Stocktaking-Tradition and Technology." *Architectural Review* 127, 756 (February 1960): 96. Reprinted in Mary Banham, Sutherland Lyall, Cedric Price, and Paul Barker eds., *A critic writes: essays by Reyner Banham* (Berkeley/Los Angeles CA: University of California Press, 1996), 49–63. In this article, Banham discussed his radical approach to technologized architecture, attempting to question the limits of the discipline. However, Jared Langevin, following Nigel Whiteley, observed that Banham's attitude seems to curiously contradict his simultaneous praise for the formal aestheticized visions and "imageability" of the work of the Archigram group. See Jared Langevin, "Reyner Banham: In Search of an Imageable, Invisible Architecture," *Architectural Theory Review* 16, 1 (2011): 2–21.
62 Banham, "A Home Is not a House," in *Rethinking Technology*, 162.
63 Ibid., 165.
64 According to Stanley Mathews, three main reasons may be relevant to why the Fun Palace was never implemented: the first reason had to do with the changes in the pattern of the local administration in the London County Council and the Greater London Council in 1964, by the Conservative government; the second reason was that the assumed indeterminacy and vagueness in both its program and form, made it hard for political authorities, possible financing bodies and the wider public, to accept and support the project; the third reason was the fact that the freedom and the improvisational strategies that were supposed to be attributed to London's working class, combined with its programmatic breaching of cultural institutions and hierarchies, made it suspect of disintegrating the social status quo. See Stanley Mathews, *From Agit-Prop to Free Space: The Architecture of Cedric Price* (London: Black Dog Publishing, 2007), 172–175.
65 Nigel Whiteley, *Reyner Banham: Historian of the Immediate Future* (Cambridge, MA: MIT Press, 2002), 212.
66 According to Stanley Mathews, situationist ideas must have influenced Price for the design of the Fun Palace, especially if we consider the common ideological and artistic roots between Price and the situationists, as well as the close friendship between Price and the situationist Scottish poet Alexander Trocchi. Although it is not clear to what extent Trocchi and Price influenced each other, and despite the fact that Price denied any inspirations for the Fun Palace from Trocchi, situationist ideas and projects (such as Constant's New Babylon and Trocchi's text *Invisible Insurrection of a Million Minds*, which described a "spontaneous university" like the Fun Palace), almost certainly affected the clarification of the aims of the Fun Palace, for both Price and his collaborator Joan Littlewood. See Stanley Mathews, "The Fun Palace as Virtual Architecture: Cedric Price and the Practices of Indeterminacy," *Journal of Architectural Education* 59, 3 (2006): 41–47. For situationists' theoretical writings see Libero Andreotti and Xavier Costa eds., *Theory of the Dérive and other Situationist Writings on the City* (Barcelona: Museu d'Art Contemporani de Barcelona, 1996). Also see Simon Sadler, *The Situationist City* (Cambridge, MA: MIT Press, 1998) for a study of the relations between situationists and urban space.
67 See Mathews, "The Fun Palace as Virtual Architecture," 40.

68 Stanley Mathews, "Cedric Price: From 'Brain Drain' to the 'Knowledge Economy'," *Architectural Design: Manmade Modular Megastructures* 76, 1 (2005): 91.
69 Mary Lou Lobsinger, "Cedric Price: An Architecture of Performance," *Daidalos* 74, (2000): 23.
70 Mathews, *From Agit-Prop to Free Space*, 13.
71 As Maddalena Scimemi explains, Price replaced conventions of representation and the functional program with a diagram of forces, in order to describe the dynamic characteristics of the Fun Palace when in use and when users activated its functions. See Maddalena Scimemi, "The Unwritten History of the Other Modernism: Architecture in Britain in the Fifties and Sixties," *Daidalos* 74, (2000): 18. Also Arata Isozaki writes that, although Price's technical drawings significantly influenced the members of the Archigram group, his visual style was radically different. Price was concerned with non-drawing and, unlike Archigram, his sketches seemed to reject formal characteristics. Instead, he used drawing to pose questions and describe methods to resolve them. See Arata Isozaki, "Erasing Architecture into the System," in *Re: CP*, eds. Cedric Price and Hans-Ulrich Obrist (Basel: Birkhäuser, 2003), 26–27.
72 Mathews, "The Fun Palace as Virtual Architecture," 41.
73 Mathews, "Cedric Price," 92–93.
74 Pask, "The Architectural Relevance of Cybernetics," 494–496.
75 Mathews, *From Agit-Prop to Free Space*, 73.
76 Ibid., 118–119.
77 Charlie Gere makes this connection between post-industrial society and second-order cybernetics. See Gere, *Digital Culture*, 116–118. The post-industrial economy is based on the economics of knowledge and information requiring more employees in the service sectors, rather than workers in industry. See Daniel Bell, *The Coming of the Post-Industrial Society* (New York: Basic Books, 2008).
78 Mathews, *From Agit-Prop to Free Space*, 75. For Pask's theory on conversation see: Gordon Pask, *Conversation, Cognition and Learning* (New York: Elsevier, 1975); Gordon Pask, *The Cybernetics of Human Learning and Performance* (London: Hutchinson Educational, 1975); Gordon Pask, *Conversation Theory, Applications in Education and Epistemology* (New York: Elsevier, 1976).
79 Yona Friedman, "The Flatwriter: choice by computer," *Progressive Architecture* 3, (March 1971): 98–101.
80 Lobsinger, "Cedric Price," 24.
81 Pask, "The Architectural Relevance of Cybernetics," 496.
82 Andrew Pickering, *The Cybernetic Brain: Sketches of Another Future* (Chicago/London: University of Chicago Press, 2009), 371.
83 Turing's machine anticipated the modern computer because it functioned by binary code, and it could resolve any mathematical problem by programming. It was an abstract finite state machine, designed to examine one of the basic questions in computer science, that is, what it means for a task to be computable, in other words the extent and limitations of what can be computed. It consisted of an infinite one-dimensional tape, divided into cells, which contained one of two possible marks – "0" or "1." It also had a writing head (like a typewriter key), which could write and erase these marks on the tape, and a reading head that could scan the current position. At any moment the next state of the machine was determined by its instructions (the transition rules), its current state, and the symbol that was currently registered. See David Barker-Plummer, "Turing Machines," *The Stanford Encyclopedia of Philosophy*, accessed January 10, 2015, http://plato.stanford.edu/archives/sum2013/entries/turing-machine. Also Gere, *Digital Culture*, 21–23. See also Turing's seminal paper "On Computable Numbers with Application to the Entscheidungsproblem," *Proceedings of the London Mathematical Society* 42, 2 (1936): 230–265.
84 Mathews, "The Fun Palace as Virtual Architecture," 42.

85 For Elizabeth Grosz's discussion of the Bergsonian and Deleuzean concept of the virtual see her book *Time Travels: Feminism, Nature, Power* (Crows Nest: Allen & Unwin, 2005). For a related discussion in the context of digital media see also Martin Lister et al., *New Media: A Critical Introduction* (London/New York: Routledge, 2003), 360–364 and the link with Gilbert Simondon's book *Du mode d' existence des objets techniques* (Paris: Aubier, 1989 [1958]).
86 Mathews explains that since the Fun Palace was originally designed with a ten-year life span, it is no surprise that Price officially declared its death on its tenth anniversary in 1976. But Price had applied the principles of the Fun Palace in two other projects: the *Oxford Corner House,* an informational leisure centre commissioned by Lyons Tea Company, and the *Inter-Action Centre* at Kentish Town, which implemented many of the participatory and social goals of the Fun Palace. The former was meant to be a new public leisure centre for the company, a sort of early cybercafé, located at the Oxford Street Corner House in London. The project, if realized, would have been an actual demonstration of the cybernetic aspects of the Fun Palace, with unprecedented computing power; it would be equipped with information retrieval systems, image-text display screens, closed-circuit television for access to educational programs and networked computers with massive storage capacity. The *Inter-Action Centre* was a multi-purpose activist community centre (commissioned by Inter-Action, a community arts organization established in 1968 by sociologist Ed Berman), built to provide services, training and creative outlets for disadvantaged people. It consisted of a simplified version of the modular structure of the Fun Palace; an open framework of standard off-the-shelf components, with modular elements that could be inserted and moved according to changes in use. As Mathews writes, although it did not share with the Fun Palace any of its technological aspects, it managed to achieve the social goals of the early project that Littlewood and Price had in mind. For more on both projects see Mathews, *From Agit-Prop to Free Space*, 177–183.
87 New Babylon was Constant Nieuwenhuys's long-lasting vision of an expandable urban network of interconnected floating megastructures, equipped with changeable architectonic elements and technological infrastructure for the development of a new nomadic and ludic culture. Conceived in the light of situationist critics of urban space, and in the context of a growing post-war economy and reconstruction period in Europe, the project comprised mobile systems and telecommunication networks, supposedly capable of realizing a labyrinthine public environment of continuous disorientation, social interaction and constantly changing artificial "unities of ambience." See Constant Nieuwenhuys's articles "Another City for Another Life," 92–95; "New Babylon," 154–169; and "The Great Game to Come," 62–63, in *Theory of the Dérive and other Situationist Writings on the City*, eds. Libero Andreotti and Xavier Costa. Barcelona: Museu d'Art Contemporani de Barcelona, 1996. Also see Mark Wigley, *Constant's New Babylon: The Hyper Architecture of Desire* (Rotterdam: Witte de With Center for Contemporary Art/010 Publishers, 1998).
88 Eleni Kalafati and Dimitris Papalexopoulos, Τάκης Χ. Ζενέτος: Ψηφιακά Οράματα και Αρχιτεκτονική (Athens: Libro, 2006), 17–27.
89 It was first presented in 1962 at the Exhibition of the Organization of Modern Housing in Athens. Also, a paper about its application to the master plan of Athens was read at the Fifth PanHellenic Architectural Congress in 1966. See Takis Zenetos, "City Planning and Electronics," *Architecture in Greece annual review* 3, (1969): 114–115.
90 Subsequent publications of the project were included in the 1970 (no. 4, 59–60), 1973 (no. 7, 112–118) and 1974 (no. 8, 122–135) issues of the Greek annual review "Architecture in Greece."
91 Zenetos, however, suggested that the existing city and its history should not be abandoned; instead it should be preserved and regenerated not as a monument

but as a living organism. See Takis Zenetos, "Town Planning and Electronics," *Architecture in Greece annual review* 8, (1974): 124.
92 Zenetos's drawing in the aforementioned Archigram issue was part of the first stage of his project (1962), when he was concerned with the creation of an artificial ground for flexible living suspended over nature with embedded supply services and electromechanical infrastructure.
93 Takis Zenetos, "City Planning and Electronics," 115–116.
94 Kalafati and Papalexopoulos, 23–24.
95 Here Zenetos's "play" recalls Constant's "homo ludens" and Price's concern with free time in the post-war leisure society. This is why Zenetos's Greek word for play ("παιχνίδι"), in his 1962 text "Town Planning and Electronics," is, we think, wrongly translated into "games" in the same text. It is obvious that he refers to play as a form of cultural and social activity, such as that in the "ludic society" imagined by Constant. See Zenetos, "Town Planning and Electronics," 122.
96 Zenetos, "Town Planning and Electronics," 122. Also Kalafati and Papalexopoulos, 46–47.
97 Even conventional service robots, says Zenetos, will be useless in this immaterial future, because developments in immaterial technological equipment will be faster and more efficient. Zenetos, "Town Planning and Electronics," 125.
98 Ibid., 124.
99 Michael Fox and Bryant Yeh, "Intelligent Kinetic Systems in Architecture," in *Managing Interactions in Smart Environments*, eds. Paddy Nixon, Gerard Lacey, and Simon Dobson (London: Springer, 2000), 91.
100 For a discussion, categories and overview, see Robert Kronenburg, *Portable Architecture* (Oxford: The Architectural Press, 1996); Tony Robbin, *Engineering a new Architecture* (New Haven CT: Yale University Press, 1996).
101 Recent texts on transformable architecture, such as Kostis Oungrinis's book, suggest many more advantages in implementing kinetic systems in architecture: from functional flexibility to dynamic control of structural loads, energy efficiency, the application of less and lighter materials, cost effectiveness and the making of future refurbishment works easier. See Konstantinos Oungrinis, *Μεταβαλλόμενη Αρχιτεκτονική: Κίνηση, Προσαρμογή, Ευελιξία* (Athens: ION publications, 2012). Kas Oosterhuis has also mentioned that kinetic systems can be used to implement active load-bearing structures, able to smartly react to changing and dynamic forces. See Kas Oosterhuis, "2006 The Octogon Interview." *ONL [Oosterhuis_Lénárd]*, accessed March 1, 2014. http://www.oosterhuis.nl/quickstart/index.php?id=453.
102 William Zuk and Roger Clark, *Kinetic Architecture* (New York: Van Nostrand Reinhold, 1970), 5.
103 Ibid., 9.
104 Richard Rogers, postscript to *Supersheds: The Architecture of Long-Span Large-Volume Buildings*, ed. Chris Wilkinson (Oxford: Butterworth-Heinemann, 1991).
105 See Fox, "Beyond Kinetic"; Fox and Yeh, "Intelligent Kinetic Systems in Architecture." Also see Michael Fox and Miles Kemp, *Interactive Architecture* (New York: Princeton Architectural Press, 2009), 13.
106 Nimish Biloria, "Introduction: Real Time Interactive Environments. A Multidisciplinary Approach towards Developing Real-Time Performative Spaces," in *Hyperbody. First Decade of Interactive Architecture*, eds. Kas Oosterhuis et al. (Heijningen: Jap Sam Books, 2012), 368–381. Also see a video at https://www.youtube.com/watch?v=e5ycPQ2Iy68.
107 See the webpage at www.display.3acomposites.com/en/products/hylite/hylite-characteristics.html.
108 The "Fluidic Muscle" (http://www.festo.com/cms/en_corp/9790.htm) is a flexible pneumatic drive developed by Festo (www.festo.com) that comprises a hollow elastomer cylinder, embedded with aramid fibers in the form of a mesh. Its initial force is 10 times higher compared with a conventional cylinder actuator of the same

size. The cylinders can shrink up to 20% of their initial length when air is pumped into them, thus working as actuators that can change the position of the nodes of the prototypes developed by the Hyperbody.
109 Biloria, "Introduction: Real Time Interactive Environments," 373. The embedded sensors and actuators are connected to the CPU through MIDI to transfer analogue data from the sensors to the processing scripts. These scripts are written in Virtools, a game software for multiple players with a plug-in developed by Festo and written in C++, used for determining interconnections between the sensor data and the expected output behavior of the prototype (by programming the output rules of the system). The state of the valves that are connected to the pistons is constantly monitored in real-time in every system unit (a Hylite panel combined with two pneumatic Festo muscles). In this way the system monitors the context of the prototype (human presence and physical state of the system). A different level pertains to the data processing scripts that work parallel to the previously acquired information about the state of the system and the sensor data, and communicate with PCI cards, which are especially used for the management of the system output, according to these scripts. These cards are programmed to receive output signals (the on/off state of the actuators) and communicate them to the system, in order to control the opening and closing of the air valves, and, thus, the actuator-driven deformations of the prototype.
110 "Muscle Room," *TUDelft,* accessed April 23, 2008, http://www.protospace.bk.tudelft.nl/live/binaries/7c13c533-6f9b-4f0b-a113-0c0ad9aaa37a/doc/index.html (site discontinued). Some credits and a brief description of the project can be found at: "Muscle Room," TUDelft, accessed August 2, 2014, http://www.bk.tudelft.nl/en/about-faculty/departments/architectural-engineering-and-technology/organisation/hyperbody/research/applied-research-projects/muscle-room.
111 Biloria, "Introduction: Real Time Interactive Environments," 376–377.
112 Ibid., 380.
113 Ibid., 98.
114 A validation of this characteristic of flexibility comes from the methodological tools that architects have developed in research projects on kinetic architecture to anticipate the function of transformable structures. For instance, Konstantinos Oungrinis reviewed many kinetic applications in architecture and engineering throughout the 20th century to determine their necessity in the contemporary social and cultural environment, and their potential capacities to respond efficiently to changing functions of architectural space. He proposed a method for collecting space-time analytical data about use and habitation, that could help designers decide whether the use of kinetic systems in buildings is necessary, as well as determine the type, extent and duration of transformability, proper materials and techniques, and the optimal solution. Oungrinis's method assumes repeated visits of the architect at the site in order to record significant qualitative and quantitative changes in the occupation of space and its conditions in different circumstances and time periods. In this way the architect could create a series of possible scenarios of actual and potential use of space, implemented in the form of space-time functional diagrams with most of the available recorded information. Oungrinis's analytical examination of the advantages and potential disadvantages of kinetic applications (compared with static architecture) was a serious research project. But the main idea that runs through his work is that architects can anticipate and predetermine the potential changes of the functional scenarios of space, as well as the way kinetic applications will respond. See Konstantinos Oungrinis, Δομική Μορφολογία και Κινητικές Κατασκευές στους Μεταβαλλόμενους Χώρους (PhD diss., Aristotelian University of Thessaloniki, 2009), 278–286.
115 Canguilhem, "Machine and Organism," 44–69.
116 Antonino Saggio, *The IT Revolution in Architecture: Thoughts on a Paradigm Shift,* trans. Stephen Jackson (New York: lulu.com, 2013), 119. Also see Antonino Saggio

"Architecture and Information Technology: A Conceptual Framework," in *The 2nd IET International Conference on Intelligent Environments (IE06)*, vol. 1, Athens, 5–6 July 2006 (London: IET publications, 2006), 9–12.
117 See Łukasz Stanek, *Henri Lefebvre on Space: Architecture, Urban Research, and the Production of Theory* (Minneapolis/London: University of Minnesota Press, 2011), 101.
118 Ibid., 102.
119 The Archigram group was criticized for a technocratic attitude that lacked a political stance. See, for instance, Mary McLeod, "Architecture and Politics in the Reagan Era: From Postmodernism to Deconstructivism," in *Architecture Theory since 1968*, ed. Michael Hays (Cambridge, MA: MIT Press, 1998), 680–702.
120 See Dimitris Papalexopoulos, *Ψηφιακός Τοπικισμός* (Athens: Libro, 2008), 47.
121 Antonino Saggio, "How," in *Behind the Scenes: Avant-Garde Techniques in Contemporary Design*, eds. Fransesco De Luca and Marco Nardini (Basel: Birkhäuser, 2002), 7.

2

INTELLIGENCE IN ARCHITECTURAL ENVIRONMENTS

Intelligence in the built environment pertains to the vision of Ambient Intelligence and ubiquitous computing. Computer scientists and engineers that are concerned with this vision aspire to create so-called *Intelligent Environments*. These environments are able to proactively respond to the needs and activities of people by means of adaptive systems, ubiquitous computing and user-friendly interfaces. Different approaches to the field of artificial intelligence (the new AI, swarm intelligence, behavioral robotics and multi-agent systems) show that the concept of intelligence is not clearly defined. By reviewing the application of intelligent systems in built space (looking at projects like the iDorm, the PlaceLab and PEIS ecology) we conclude that, although these embedded systems are capable of learning, activity recognition and proactive anticipation of users' needs (which somehow revisits the proposals of the experimental architects of the 1960s), they essentially respond to already preconditioned preferences through automated rule-driven services. Interesting counter-examples come from the so-called *end-user driven* systems, namely environments that enable user involvement in their functionality and promise a future for more open-ended spaces.

Artificial Intelligence – From Symbolic AI to Emergent Multi-agent Systems

The *Generator*, one of the first attempts to apply artificial intelligence techniques in architecture, was a proposal for a retreat center at the White Oak Plantation on the coastal Georgia–Florida border, designed by Cedric Price in collaboration with John Frazer from 1976 to 1979. These architects conceived the project as a kit-of-parts of 150 12´ × 12´ moveable cubes (distributed on a permanent grid of foundation pads) and other mobile elements, such as catwalks, gangways and sliding screens. If the project was constructed, its residents would be able to

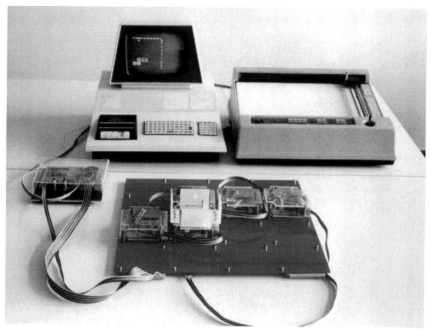

FIGURE 2.1 *Generator*: view of working electronic model between 1976 and 1979, Cedric Price, architect, chromogenic color print mounted on cardboard, 12.6 × 17.3 cm. Cedric Price Fonds, Collection Centre Canadien d'Architecture/Canadian Centre for Architecture, Montréal.

determine its physical layout by requesting changes that would be implemented by a mobile crane able to relocate the cubes accordingly. But at the same time, the building would be able to make decisions and suggest changes to reorganize itself. Each cube would be equipped with an embedded single-chip microprocessor, able to communicate its position to the central computer, which would keep an archive of its past configurations (Figures 2.1 and 2.2). If the building remained inactive for a long period of time, it would register its own "boredom," and then it would propose new configurations.[1] Thus, it could demonstrate a proactive and anticipatory behavior, performing autonomous, self-generated decisions which, as we shall see later, are some of the central concerns of Ambient Intelligence. As John Frazer has explained, the Generator would work for the users' benefit, gradually improving itself, because it would be able to "…learn from the alterations it made to its own organization and coach itself to make better suggestions."[2]

Although we can use the broad concept of "intelligence" to characterize the Generator, and despite the fact that it was considered to be one of the first "intelligent" buildings,[3] intelligence and its use in relation to buildings certainly needs clarification. Therefore, in the following section, we will look into the field of artificial intelligence and later into ambient intelligence, to examine the potential application of intelligence in the built environment.

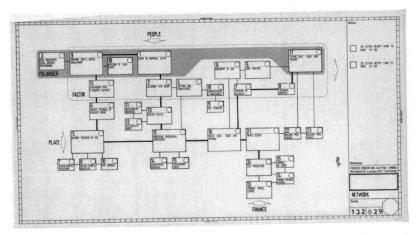

FIGURE 2.2 *Generator*: initial design network showing three starting points, Cedric Price, 1976–1979, black ink on paper, 37.9 × 71.5 cm. Cedric Price Fonds, Collection Centre Canadien d'Architecture/Canadian Centre for Architecture, Montréal.

First Steps in AI: The Symbolic Approach and Connectionism

The idea that intelligence could be mechanized and reproduced by artificial means, can be traced back to the 17th century (see also Chapter 3).[4] Yet, it was mathematician Alan Turing who systematized this possibility in the 20th century with his hypothetical Turing Machine, as well as the so-called Turing Test.[5] The prospect of artificial intelligence was enhanced after the second half of the 20th century by the development of the formal theory of computation, the construction of electronic stored-program computers and the discovery of the neuron. But it was officially formulated as a scientific discipline at a conference in Dartmouth in 1956, where the term AI, suggested by the conference organizer John McCarthy, was ratified. Although AI had its own techniques and concepts, it was not an isolated scientific discipline in the study of intelligence; it drew methods from other areas and was part of a wider multi-disciplinary field of studies, including the philosophy of mind, the philosophy of language and cognitive psychology, all studying intelligence in both natural and artificial systems, and concerned with a possible theory of mind.

Research in AI followed two main directions, although they were both historically and philosophically rooted in a seminal paper written by Warren McCulloch and Walter Pitts in 1943.[6] The first direction in AI, the classic approach (aka "Good Old Fashioned AI"), prevailed roughly between 1950 and 1980. Allen Newell and Herbert Simon expounded this approach by developing the General Problem Solver, a computer program based on the analysis and coding of human thinking into computational rules.[7] Newel and Simon along with other researchers of this early AI trend, considered intelligence to be a function of informational or symbol-processing machines – either computers or human brains. The main idea, which seemed to encompass a critical break with early cybernetics as well, was that computers could be programmed to simulate a physical symbol (or formal) system, namely a system of physical patterns (symbols) that could be combined with other patterns into structures (expressions). These structures could be further manipulated by certain processes.[8]

Early cyberneticists, as we saw in Chapter 1, worked with electromechanical machines rather than programming, and did not participate, as John Johnston explains, in the shift from hardware to software that characterized early AI.[9] While cyberneticists emphasized concrete embodiment and performativity by constructing machines that demonstrated signs of adaptive and intelligent behavior, symbolic AI was founded on the hypothesis that cognitive processes are disembodied, abstract and symbolic. This approach evolved in parallel with the computational paradigm in psychology – the information processing model in cognitive science called "cognitivism." This further developed the idea of the brain as an information processing device, able to manipulate abstract data structures and physical symbols, like computers, which was sufficient to assume general intelligent action.[10]

The second major branch of AI was "connectionism," an alternative model of computation and cognition that emerged while symbolic AI was developing. The idea was initially instantiated in a learning algorithm, the "perceptron" (later implemented in a custom-built machine), invented by Frank Rosenblatt in 1957. It consisted of excitatory and inhibitory links (with variable strengths) between simple agents (neurons able to fire according to input and the state of the neighboring neurons) that would progressively modify their connections, resulting in the emergence of intelligence through learning. Although connectionism went through a 15-year period of lack of interest by scientists and funding cuts from the US Defense Department (from 1970 and on),[11] it revived in the mid-1980s, after scientists acknowledged dead ends in symbolic AI, and perhaps because it fitted in a larger cultural, societal, philosophical and scientific shift favoring decentralized structures. Proponents of connectionism stressed the idea that intelligence is a function of "connections" between several processing units, and envisaged machines able to learn by experience by means of a system of independent and interacting computational agents. Inspired by the neural organization of brains, they implemented their systems by many types of parallel distributed processing systems and artificial neural networks.[12] The advantages of the connectionist against the classical approach to artificial intelligence, was its non-deterministic view of intelligence and the potential to perform operations, such as pattern recognition, that could not be achieved by programming in the classical sense. Instead of programming large amounts of specific rules managed by a central processing unit (as in classic AI), connectionism favored decentralized control, parallel distributed processing and adaptive feedback.

Proponents of both the symbolic and the connectionist approach to AI, assimilated intelligence with the cognitive processes of the human mind, such as thinking, reasoning, inferring, deducing, understanding, and learning, leading to actions and behaviors that would require intelligence if done by humans.[13] But this approach had many difficulties for researchers and AI theorists, because it assumed that the problem of AI was a problem of simulation, despite the obvious discrepancies between human and artificial intelligence. While AI machines performed complex rapid calculations strikingly superseding brains, AI was problematic in relation to human common sense, the imperfection of human thinking and reasoning, the role of emotion in cognition and the role of the material substrate of intelligence.[14]

Stuart Russel and Peter Norvig, reviewing all the different definitions of intelligence in the context of AI, proposed a way to categorize them by distinguishing between human and rational intelligence, as well as intelligence related to thinking and that related to action and behavior. Thus, they distinguished between intelligent systems that can either think or act like humans, and those that can think or act rationally (according to idealized models). However, they argued that AI researchers should not try to simulate human intelligence because modeling biological features in engineering is not always the right method (as was the case with the design of airplanes). Rather, artificial intelligence should focus on rational system behaviors.[15]

Artificial Intelligence and the Built Environment

To explore the role of intelligence in the built environment, we should start by addressing the concept within the discipline of AI. But in the light of Russel and Norvig's suggestion, as we shall see later, an intelligent architectural environment does not act or think like humans but is rather capable of rational autonomous action, providing the appropriate services in a given situation, and optimizing the functions of the environment.

Classical applications of AI operate well in environments that need high computational power and data storage (such as in chess playing), or in places where efficiency depends on automation, specialized routine operations and control (as in industrial robots). They are also useful in computational problem-solving, as in expert or knowledge-based systems.[16] In this case they contribute directly to a large number of creative and work-related activities, such as consultation and diagnostic operations in medical environments, as well as health monitoring and control for the elderly in domestic environments.[17] However, these applications can hardly operate in everyday situations where events are indeterminate, unaccountable and depend on the complexity and context of human activities.[18]

On the other hand, connectionist applications of AI seem to be more promising for architectural environments due to their learning and adaptive capacity. From an architectural standpoint, Maria Luisa Palumbo proposed the potential use of artificial neural networks in architecture, assuming that the classical model of AI, which follows an analytical logical direction, is not sufficient for intelligent environments and should rather be replaced by a holistic connectionist strategy. This would make use of integrated neural networks with learning capacities, functional flexibility, adaptation and other characteristics of living beings.[19] As we will see later in the chapter, the learning capacity has been a central concern for the design and implementation of so-called *intelligent environments*. The *Adaptive Home* for instance, a rather early example of an intelligent house, was able to observe the reactions of its occupants to changing operational conditions and predict their needs through acquired experience. The house would use predictive neural networks and reinforcement learning for efficient management of its energy supply systems (lighting, heating and ventilation) and comfort levels.[20]

But besides learning, Palumbo's ideas included embodied machines and distributed de-centralized systems, thus pointing to a relatively recent trend in the philosophy of mind and cognitive science that rejects Cartesian dualism and emphasizes the embodied nature of intelligence. This trend in cognitive science was expounded by Francisco Varella and his theory of "enaction,"[21] and was physically implemented in robotics by computer scientist Rodney Brooks.

Embodied Cognition and the New AI

The shift in the understanding of intelligence proposed by the researchers of the new AI is a larger inter-disciplinary theme in linguistics,[22] cognitive science,[23] philosophy of mind,[24] robotics[25] and cognitive neuroscience.[26] In this context, mental representations and cognitive processes are not symbolic and disembodied, as in a classical cognitivist view that favors a Cartesian dualist attitude to the mind–body relation, but rather embodied and situated, emerging out of the constraints of the body and its sensory-motor interactions with the environment.[27] In robotics, until the end of the 1980s, the main trend was founded on a Cartesian model in Western epistemology, which emphasized centralized control, reflecting what Kevin Kelly considered to be the first type of machines that can be constructed, that is, mechanical predictable and hierarchical systems, such as the clock, following linear and sequential operations commanded by a central authority.[28] This was featured in industrial robotic applications, which involved precise and stable operations guided by a particular albeit flexible software. Service robots, on the other hand, which would inhabit the social environment and cater for human needs, seemed incapable of coping with the uncertainty and fuzziness of everyday spaces.[29]

By the end of the 1980s, Rodney Brooks had designed and constructed robots that would directly interact with their environment through sensory-motor operations, without central control and pre-programmed representations (model). Inspired by the behavior of simple primitive organisms, these physical machines were equipped with a distributed system of sensors and actuators, and functioned by prioritizing actions according to a hierarchical set of in-built behavioral modes (starting from simple and moving on to more complex tasks) – what Brooks called

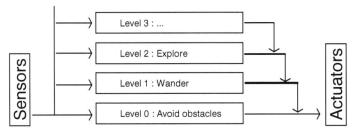

FIGURE 2.3 Diagram of subsumption architecture

"subsumption architecture" configurations (Figure 2.3).[30] Thus, intelligence was not an aspect of logical calculations on the basis of pre-given representations as in symbolic AI, but an emergent epiphenomenon of machines that could interact with local situations in the environment in which they were embedded, constantly adjusting their behavior by suppressing actions of lower priority. Intelligence would thus be, as Brooks stated, "in the eye of the observer."[31]

Such a non-representational approach to AI would be more appropriate for the unpredictability and the fuzziness of everyday environments. It would be able to deal directly with it, and allow intelligence to emerge from those interactions between the physical structure and its dynamic content.[32] For instance, computer engineers have used "subsumption" architecture, as defined by Brooks, to implement user-driven intelligent environments that can manage unpredictable situations (which we will discuss in Chapter 5).[33] Also, in line with Brooks's subsumption architecture in robotics, architect Konstantinos Oungrinis's initial research in kinetic architecture investigated the idea of architectural environments that can sense and respond on the basis of functional priorities. The functionality of such decentralized systems would emerge from simple environmental interactions. Sensory-motor activity, analysis and evaluation of successful interactions, would make the system learn and gain experience in order to optimize its prioritized tasks, such as energy efficiency and adaptive structural stability. Subsumption architecture would subsume and override other hierarchically lower or conflicting actions, such as functional flexibility and user desires.[34] But the new AI is also characterized by a wider shift in the understanding of intelligence in artificial systems, and a developing concern for multi-agent and collective distributed systems, able to demonstrate emergent swarm intelligence, which marks a break with the centrality of Varela's and Brooks's embodied subject.[35]

Swarm Intelligence

The study of swarms comes from the work of entomologist William Morton Wheeler (1911) and zoologist Pierre-Paul Grassé (1957), as well as the work of 19th-century research in ethology, a discipline that combined zoology, comparative anatomy and the study of ecosystems, to understand the way living organisms interact.[36] But in the last 25 years, the conjunction of ethology with computer science (including input from Artificial Life in software, evolutionary robotics, complex adaptive systems and cellular automata), contributed to the understanding and computational modeling of the collective intelligence and synchronized behavior of biological swarms (such as schools of fish or flocks of birds),[37] and led the way for the subsequent employment of swarm intelligence in the construction of multi-agent artificial systems (such as multi-robot systems and multi-agent simulations).[38] Swarms demonstrate collective, decentralized overall behavior, accomplished by local interactions between individual entities that can self-organize and coordinate on the basis of simple behavioral rules.[39] Besides the assumption that there is no central control, a precondition for swarms is that collectivities assemble in a given concrete environment. In this environment, units are distributed by connectivity (a status provided by technological or political means), that is, the ability to exchange information through a common set of terms that holds this topological configuration together. This is what separates swarms as forms of grouping from other collective phenomena, such as multitudes (which comprise individuals with their own intentions and desires that can exist outside the collectivity), or electronic networks that do not necessarily imply collectivity and aggregation in physical spaces.[40]

Kas Oosterhuis has written about swarm intelligence and its possible association to architecture. He says that the understanding of swarm intelligence as emergent behavior of complex interactions between simple components, can be applied in the conceptualization of "a new kind of building," as he calls his Hyperbodies: "It seems possible to apply the same definition of intelligence… to the relations that are built (both in the design process and in the actual operation of the construct) between all actuators/components assembled into a building."[41] He thinks that the nodes of the kinetic mesh structure (the points of intersection of the actuators), other interconnected buildings and the users (through their local movement or information transmitted through the internet), are all different intercommunicating swarms capable of influencing the form and content of the structure (for instance by projecting images on its interior), sometimes in unpredictable ways.[42] As Oosterhuis further explains, projects like the *Muscle NSA* (a constructed prototype of the speculative Trans_PORTs 2001), can calculate in real time the positions of its actuators in relation to their neighbors. These actuators can interact with each other (without having knowledge of the overall form of the structure), and conform to three simple rules that adjust their distance, speed and direction: "don't come too close to your neighbors," "adjust your speed to the neighbor's speed" and "try to move towards the center."[43]

Muscle NSA was exhibited at the Venice Biennale in 2000, and at the Centre Pompidou in 2003 for the Non-Standard Architecture Exhibition. It was a soft volume wrapped in a mesh of 72 Festo muscle tensile pneumatic actuators, capable of orchestrated and cooperative length changes that would result in modifying the overall shape. This involved contracting and relaxing motions of the actuators (produced by varying the air pressure pumped into them), and constantly calculating the relative positions of the nodes of the mesh, thus behaving like a swarm. As Chris Kievid and Kas Oosterhuis explain: "They all work together like a flocking swarm of filaments in a muscular bundle each performing in real time."[44] A number of pressure, motion and proximity sensors were attached on the skin of the exhibited prototype to register the motion, distance and behavior of visitors. The software control system used numerical data from these sensors as input to activate the structure; it would process the 24 digital MIDI sensor signals and the 72 actuator signals in the form of digit strings, constantly updating the variable parameters of a number of scripts, namely behavior states running in real-time. These states were presets, selected according to the internal state of the system, and could be combined with local user interferences to produce a totally unpredictable outcome in the form and motion of the structure (shaking, vibrating, bending or twisting).[45]

Although Oosterhuis argued that this "quick and dirty prototype"[46] should be able to adapt to a variety of uses (such as performance space, disco room, TV room, research room, reception room etc.), the potential for transformation was rather restricted by the physical limits of its actuators and skin. This must have prevented it from demonstrating the apparent formal dynamism and flexibility of swarms. It is also not clear whether the designers actually used a swarm optimization algorithm to animate the structure, and whether the association with swarms was simply a metaphor. Furthermore, the emphasis on the performative aspects of its external form and surrounding space, made it seem more like an interactive installation or kinetic sculpture, as Kari Jormakka argued,[47] constructed to demonstrate the emerging potential of information-driven structures and environments that are sensitive to the activities of people.

What is interesting in Oosterhuis's attempt to connect architecture with swarms, is that the system he imagines, comprises heterogeneous parts, both technological and social agents (structural components as well as local and internet users), interacting with each other in real-time and in physical space. He imagines an architecture "where all players (including all building materials) are seen as potential senders, processors and receivers of information, and where all players communicate with members of their own flock and with other flocks in real time."[48] Although this hypothesis does not conform to one of the most important properties of swarms, that they comprise simple, primitive and non-intelligent components, the idea and potential of a heterogeneous assemblage of social and technological agents will be examined in Chapter 4, in the context of human–computer interaction and the social theories of technology.

Several other applications of swarms in design can also be considered, although they all tend to be speculative visions rather than implemented physical projects. For instance, Roland Snooks and Robert Stuart-Smith have explored swarm-based generative design methodologies for architecture and urban design, employing concepts such non-linearity, self-organization, emergence and dynamical systems theory in design (www.kokkugia.com). Pablo Miranda Carranza has also considered a similar approach with his experimental work configurations (Figure 2.4). He explored the possibility of producing computing

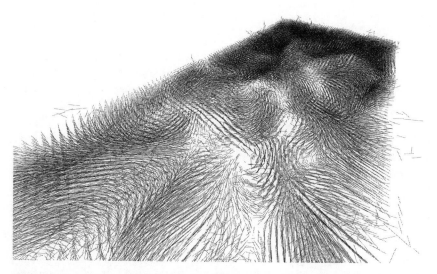

FIGURE 2.4 *Sematectonic Lattice*, Pablo Miranda Carranza, 2000. Courtesy Pablo Miranda Carranza.

and computed architecture by means of algorithmic techniques performing simple calculations like swarms.[49] These applications, however, are rather concerned with digital design processes, rather than actual implementations of swarming systems in physical environments and structures, like Oosterhuis's projects.

Computer scientists, on the other hand, have constructed physical implementations of swarm intelligence in robotic systems. Inspired by biological swarms, they have created self-organizing, multi-agent robotic systems that, compared with Brooks's behavior-based robots, are structurally more flexible and functionally more adaptive. Research programs such as the *Swarm-bots*, funded by the Future and Emerging Technologies program of the European Commission, or a project undertaken at the Multi-Robot Systems Lab at Rice University, are developing multi-robots that consist of autonomous sensor-equipped units (the s-bots), which use swarm simulation control software and communication protocols.[50] Through coordinated motion and self-assembly, these robots are able to efficiently perform cooperative tasks that cannot be achieved by a single robot (such as avoidance of obstacles, passing over a hole, motion on a rough terrain, transporting bodies in extreme conditions, or domestic activities such as cleaning).[51]

In the context of architecture, a direct speculative application of the cooperative skills of swarm-bots would be the construction of buildings by modules that are themselves robotic swarm agents, able to reconfigure the overall form according to changing circumstances. The author explored this idea of cooperating robotic modules in a speculative design project called *Swarm-Roof*, a reconfigurable shading system consisting of a group of self-organizing swarm robot discs embedded in a roof structure configuration (Figures 2.5 to 2.7). By changing

Intelligence in Architectural Environments **65**

FIGURE 2.5 Equidistribution, *Swarm-Roof*, 2009

FIGURE 2.6 Self-assembly of one group, *Swarm-Roof,* 2009

FIGURE 2.7 Self-assembly of two groups, *Swarm-Roof*, 2009

its overall organization according to user commands (such as "distribute evenly across the surface of the box," "assemble to a point," or "create several assembly points"), the group of discs could adjust its shape and density, thus projecting different shadow patterns on the ground.[52]

But a much more radical project in this line of research comes from the work of Fabio Gramazio, Matthias Kohler and Raffaello D'Andrea and their *Flight Assembled Architecture* installation (2011) (Figures 2.8 and 2.9). In this installation (which performed for a period of four days), four flying robots, the "quadrocopters," carried 150 foam bricks, one by one, to build a 6-meter high physical structure.[53] The robots incorporated a specialized gripper made of three servo-powered pins able to puncture and hold the brick configurations (Figures 2.10 and 2.11). Through feedback from the robots, a subsystem would coordinate the machines, plan their trajectories, and manage the tolerances of the pick-up, the placement of bricks, the battery recharge control and the flight speed. The robots used two predetermined freeways around the structure for their flights, and a space reservation system to avoid collisions; once a freeway was cleared, it could be reserved by one of the machines to grant its access before executing its flight.[54]

Intelligence in Architectural Environments **67**

FIGURE 2.8 *Flight Assembled Architecture,* Gramazio & Kohler and Raffaello D'Andrea in cooperation with ETH Zurich, FRAC Centre Orléans, France, 2011–2012. © François Lauginie.

FIGURE 2.9 *Flight Assembled Architecture,* Gramazio & Kohler and Raffaello D'Andrea in cooperation with ETH Zurich, FRAC Centre Orléans, France, 2011–2012. © François Lauginie.

68 Intelligence in Architectural Environments

FIGURE 2.10 *Flight Assembled Architecture,* Gramazio & Kohler and Raffaello D'Andrea in cooperation with ETH Zurich, FRAC Centre Orléans, France, 2011–2012. © Gramazio & Kohler.

FIGURE 2.11 *Flight Assembled Architecture,* Gramazio & Kohler and Raffaello D'Andrea in cooperation with ETH Zurich, FRAC Centre Orléans, France, 2011–2012. © François Lauginie.

From Artificial Intelligence to Ambient Intelligence

Having examined the capacities of AI systems for architectural space it is now time to focus on specific applications of information technology in architectural space. *Ambient Intelligence* (AmI) is a research program in computer science concerned with the creation of *Intelligent Environments* (IEs).[55]

Intelligent Environments

Intelligent Environments are computationally augmented spaces, engineered to respond to the presence and activities of people in an adaptive, proactive and discreet way, supporting and enhancing human life. Three key technologies can implement this AmI vision: Ubiquitous Computing, the idea of computers coming out in the physical world (by integrating microprocessors and sensor-effector devices into everyday things), first conceived by Mark Weiser in 1988;[56] Ubiquitous Communication, which facilitates communication between objects and users through wireless networks; and Intelligent User Interfaces, which allow users to control the environment and interact through context-dependent and personalized voice or gesture signals.[57] Although research in this area addresses several types of environments (cars, workplaces, classrooms, factories, clothing, and even online virtual worlds), we will focus on domestic space, as discussed by AmI researchers, who gather every year in international meetings, such as the International Conference on Intelligent Environments (organized by the Institute of Engineering and Technology and the Association for the Advancement of Intelligent Environments).[58]

Houses with automated functions have been around since the 1950s and 1960s; for instance General Electric's *All Electric House*, built at Kansas in 1953, involved remote controlled assistive services, such as turning on/off the lights when the time was right, watering the garden or making coffee at a specified time. Since then, however, AmI researchers have developed techniques to enhance and automate domestic habitation and services, by assessing situations and human needs to optimize control and performance in architectural space. Intelligent environments use environmental information acquired through human activity recognition systems. They use this information as feedback to obtain knowledge and experience, through learning, memory and proactive anticipation mechanisms, in order to produce rules to adapt and respond accordingly to personalized needs and changing behaviors.[59] Their internal processes use AI systems, neural networks, fuzzy logic and intelligent agents, in order to optimize the way the environment operates, despite uncertainty. A common database processes environmental information, provided by a sensor network, in relation to preset targets to determine the state of the environment and the behaviors of users. Based on these deductions, the system uses decision algorithms to perform a chosen action using actuators, thus changing the state of the environment configurations (Figure 2.12). As Diane Cook and Sajal

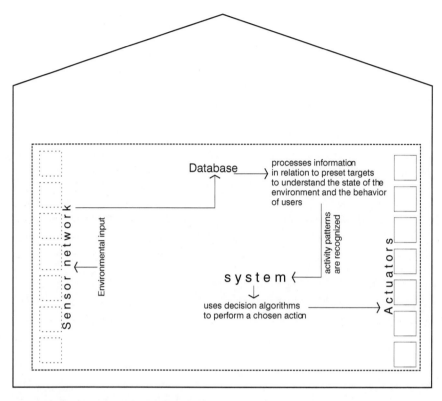

FIGURE 2.12 Diagram of Intelligent Environment

Das explain, the increasing intelligence of these environments is due to the developments of sensor–actuator technologies, the human activity recognition systems, as well as the learning and decision-making systems.[60] But the question that presents itself is what actually makes these environments intelligent – given the difficulty in defining intelligence in the context of AI (earlier we saw the problems in determining the meaning of intelligence in artificial systems).

When engineers and researchers use the term "intelligent" in the context of AmI, they mean something more than "automated," because certain properties of contemporary intelligent environments, like learning and proactivity, were not part of early automated houses. Moreover, there is a feeling that the term "intelligent" points to something higher than the term "smart," at least judging from the way the terms are used in everyday language.[61] Harvard professors Michelle Addington and Daniel Schodek argue that "smart" environments can perform simple operations, like collecting environmental information by detecting and monitoring the building, whereas "intelligent" environments would use that information as feedback to acquire knowledge and experience to optimize control and efficiency.[62] Jari Ahola in the 2001 issue of *ERCIM News* (the magazine of the European Research Consortium for Informatics and Mathematics), explained that intelligence is not simply the computational and communicative capacity of a system, but its ability to learn and be informed about the constantly changing needs of users. This does not only mean a system that thinks, but also a system able to learn how to respond and operate efficiently, despite the changes in user preferences or the layout of spaces.[63] Therefore, the term "smart" refers only to some of the capacities of what we call "intelligent"; the latter, in the context of AmI, implies developing cognitive properties such as learning, reasoning, making inferences, as well as evaluation and reflection.[64]

Architects like Ted Krueger and Ranulph Glanville, however, have objected to the idea that intelligence is the capacity of buildings to provide autonomous and automated services. Instead, intelligence in architectural spaces should be associated with the capacity of the system to interact with people. Intelligence, in this context, is misused, because intelligent buildings do not provide interactive and participatory services, but only respond and react to stimuli.[65] In Chapter 4 we will propose a different understanding of intelligent environments; rather than smart service providers, we can conceive them as active, conversational partners.

Functional Capacities

As already mentioned, IEs can use expert systems and fuzzy logic to support habitation by managing environmental information and provide consultation and directives for users; the users, though, are in charge of the final decision in activating certain behavioral modes in the environment. Besides consultation and acting only under human supervision, in certain unpredictable circumstances, IEs can take initiative and actually make their own decisions, thus reducing the presence of human supervision. By means of integrated thermostats, lights, ventilators, shading systems and security devices, IEs can detect, monitor and optimize the environmental state of spaces.[66] An air quality control and management system for instance, was developed for the PlaceLab, an intelligent apartment of the House_n research program at MIT. Also Grimshaw Architects' Rolls-Royce headquarters and manufacturing unit building (2003) uses moveable facade blinds that can react to the changes of the sun position, in order to anticipate and optimize the intensity of natural light in the interior space (Figure 2.13).[67]

Another capacity of IEs is the anticipation of human needs and desires by proactively recognizing a complex situation or event and respond accordingly. This is the subject of a research project under the title "proactive computing," undertaken by corporations like Intel and IBM. Intelligent environments like the *Adaptive House*, the *PlaceLab* and the *MavHome*, use techniques for the recognition of patterns of human activities in order to anticipate as precisely as possible the future positions, routes and activities of their occupants.[68] In this way, they can automate and optimize control of devices in the environment, determine anomalies and lower their maintenance costs and resource consumption. They can also provide special health advantages and services for the elderly and handicapped people.

FIGURE 2.13 Day view looking into production line, Grimshaw, Rolls-Royce Manufacturing Plant and Headquarters, Chichester, West Sussex, United Kingdom, 2003. © Edmund Sumner / VIEW.

On a more advanced level, the environments can selectively respond to certain situations by deciding when and where an action should be taken. For instance, they can cool or heat only the local immediate space of certain users instead of the whole room, thus optimizing environmental performance.[69] This is also evident in building facades that use environmentally sensitive skins, which only respond selectively and locally by adjusting their state according to the particular conditions of the interior space.[70]

In general, IEs can operate autonomously, automating selectively and sometimes proactively services and functions in space, providing consultation when needed, and catering for the elderly and people with special needs. By recognizing and learning human behavior patterns they can evaluate situations and needs in order to optimize their performance. But as we will see, IEs cannot recognize human behaviors in complex situations and, therefore, sometimes they cannot respond proactively to human needs.

Activity Recognition in IEs

Intelligent environments can hardly understand and recognize human desires and intentions, because human behavior is context-dependent and its meaning is ambivalent. John Thackara, the founder of the *Doors of Perception* design network, has argued that the vision of ambient intelligence for ubiquitous, unobtrusive and adaptive systems, able to respond intelligently and proactively to our desires, is based on a mistaken hypothesis about the capacities of software systems.[71] Yet, there have been several attempts, in the context of activity recognition, to develop systems able to recognize – to a certain extent – human activities, through context sensing.

Context sensing is the capacity to perceive and evaluate the environment in which a human activity takes place, by processing information from many sensors distributed in several locations in space and in objects in space.[72] Increasing the amount of sensor information reduces the error that may occur in activity recognition, although this is restricted within a predefined range of activities. The system is initially built with a finite number of specific rules that determine the relations between sensor-input and activity recognition. Yet, these rules are constantly modified as the system observes the actual results in space, adapts through learning, and produces new rules from the acquired experience.[73] This is the actual operation of *iDorm*, described in the next section.

iDorm

The iDorm is an experimental project for an intelligent student apartment, developed by the Intelligent Inhabited Environments Group at Essex University.[74] It is a platform for the study of the potential of AmI to monitor, modify and improve the spatial conditions (ventilation, heating, lighting, sunniness and security) of domestic space, according to the changing needs of occupants. The room has a network of temperature, humidity, light and motion sensors, a security system and a camera that monitors the room from the outside. This network is connected through communication channels to several actuator devices: ventilators, air-heaters, door lock mechanisms, mobile vertical blinds, automated window opening systems, and a light dimmer. There are also components that are attached on the furniture, and can record and learn the behavior of the users, in order to provide services that can improve the quality of their life. Using fuzzy logic, the environment can model observed activities by recording and interpreting the interactions of the users with these devices (window blinds, light dimmers etc.) to create operational rules. At any time users can override the agent that controls these operations by manually adjusting the devices according to their preferences. These changes are then registered by the agent. If necessary, the agent can modify, erase or add new rules, in order to adapt its function to the changing behaviors of users, and respond proactively to their needs at any time.[75]

In particular, researchers performed a 5-day experiment to train an Adaptive Online Fuzzy Inference System Agent (AOFIS) with a Fuzzy Logic Controller (FLC), which could learn and adapt to the needs of a single user who stayed in the room during this time. During the first 3 days – the learning period – the agent would observe and register the user's interactions with devices and furniture items, in order to model his/her behavior and generate operational rules. The room used seven light and temperature sensors, pressure sensors attached on the chair and the bed, as well as ten actuators, dimmer spotlights, window blinds, heater, word processor and media player.[76] During the next 2 days, the agent would run in its normal mode, in order to evaluate the ability of the FLC to manage the environment according to the needs of the user by monitoring the conditions of the devices, using the generated rules (from the initial learning period). On the other hand, the user could override the operation of these devices by changing the adjustments of their control mechanism and modifying the current responses of the agent. The agent would then register an input–output snapshot (the state of sensors and actuators with the new values of the latter as adjusted by the user) and feed it back to the rule adaptation routine. Thus, it would change its rules or add new ones (although the old ones were retained for future reference), as a result of the new preferences and changed behavior of the user (a repeated behavior meant a significant change in the user's habitual behavior).[77] Thus, the environment could sufficiently model the behavior of the user through learning, and discreetly adapt to his/her preferences, by extending, changing or adding new operational rules as a result of the changes made by the user.[78]

Although the iDorm presents significant capacities for activity recognition, in general there are some important restrictions. In fact recognition is most effective only for singular activities, executed one after the other. Difficulties arise when users perform concurrent activities, or when two or more residents interact to perform an activity, although researchers have explored attempts to overcome these problems.[79] This is a research challenge:

> [t]he modeling not only of multiple independent inhabitants but also the accounting for inhabitant interactions and conflicting goals is a very difficult task, and one that must be addressed in order to make smart environment technologies viable for the general population.[80]

Another challenge is the recognition of intention; what users intend to do, not what they already do, which would allow the system to be proactive and anticipate future user preferences. To resolve this, researchers have suggested techniques to determine behavioral patterns from an analysis of everyday routine activities and habits, despite the fact that these patterns may change later or demonstrate a degree of uncertainty.[81] The following project was an experiment for the study of activity recognition and the related problems.

House_n: PlaceLab

House_n is a multi-disciplinary research consortium at the MIT department of Architecture, funded by companies like Compaq, Proctor & Gamble and Superior Essex.[82] Researchers have undertaken several projects to study how new technologies, materials and strategies in design can lead to dynamic and evolving spaces that are responsive to the complexity of life. In the context of House_n, the *PlaceLab* is an experimental platform for the study of intelligent domestic space, embedded within an existing house close to MIT, where researchers could systematically evaluate strategies and technologies in a real-world setting.[83] Throughout the experiments, which lasted for a few days or weeks, a sensor network would monitor and register the detailed activities of volunteer participants (in different and complex situations), in order to infer specific data for real situations and human activities, in a place that provides the conditions and the infrastructure for "free habitation."[84]

The house contained 15 prefabricated components, equipped with a microprocessor and a network of 25 to 30 replaceable sensors, able to detect changes in the state of appliances and furniture, movements of people, changes in temperature and lighting conditions, as well as monitor the function and flow of energy and water supply networks. The motion of objects was detected by small detachable wireless motion sensors, the "MITes," which could be relocated and attached onto different spots in the house. In addition, nine infrared and color cameras, and 18 microphones were distributed around several locations in the apartment. Twenty computers used image-processing algorithms to select those images from four videos and the associated sound signals that best recorded the behavior of an inhabitant. These signals were then synchronized with all the other data from the sensors, and were stored in one single mobile hard drive. Researchers could record activities within a controlled environment, and therefore they were able to carefully study how residents reacted to new devices, systems and architectural design strategies in the complex conditions of a house. By contrast to other similar platforms, the polymorphous and ubiquitous sensor infrastructure could provide a single, unified and synchronized set of data to the researcher.

In particular, researchers undertook a series of experiments based on 104 hours of activities of a married couple, who lived voluntarily in the PlaceLab for 10 weeks. The aim was to determine human activity recognition algorithms, and the kind and amount of required sensors. What was noteworthy in that experiment was that the users did not know the researchers and the way the sensors functioned and, therefore, they were not inclined to adjust their behavior according to the needs of the experiment. Their long-term stay and the ordinary form of the interior, made them behave naturally because they were not constantly aware that they lived in an experimental setting.[85]

In the particular experiment, the house was equipped with more than 900 sensors of different types (including infrared motion detectors and RFID tags), attached onto pieces of furniture and devices, either immobile or mobile, such as doors and closets, remote controls, appliances, cutlery and even food. The users had to wear RFID tags on a wrist bracelet, as well as body posture sensors.[86] The human activities that took place in their respective room (eating in the kitchen), were correctly recognized. But the system could not recognize those activities that took place in areas other than the prescribed ones, such as eating away from the kitchen table or brushing teeth while moving around the house. The researchers observed similar problems when activities were interrupted by other activities, when multiple activities happened concurrently, or when the couple performed activities both collaboratively and separately. Also, in many cases the RFIDs proved incapable of tracing those objects that were moved by the inhabitants.[87] The experiment showed that for the system to recognize activities successfully, it should have information from ubiquitous sensors, placed all over the environment, on objects, furniture and people, but those activities should not be too complicated. This means that activity recognition systems may be acceptable for people with special needs, but they would have trouble coping with the complexity of everyday situations in most houses.

The PlaceLab is one of several intelligent environments that approximate the idea of an intelligent servant, able to recognize people's activities and needs, in order to respond immediately and proactively. These environments assist users by simplifying and anticipating their everyday domestic activities; at the same time they manage efficiently and economically the energy resources, namely lighting, heating and ventilation systems (building management systems and integrated building systems), and control and security services. Following this idea, the next project employs robotic assistants informed by a network of distributed sensors within the domestic environment.

PEIS Ecology

The *PEIS Ecology* project was carried out at the AASS Cognitive Robotics Systems laboratory at Orebro University, Sweden. It was an intelligent environment that used robots, and other distributed sensor and effector devices, called PEIS (physically embedded intelligent systems), for the provision and execution of assistive services in domestic spaces, such as the transfer of objects to handicapped users or floor cleaning. These robots, which were capable of locomotion, face recognition and the execution of simple actions, did not operate in isolation; the system could be rather conceptualized as an "ecology" of communicating and cooperating simple components, i.e. the robots and the distributed sensors in the house. Therefore, the way the project functioned did not derive from the individual capabilities of the PEIS components, but emerged from their interaction and cooperation.

Every subsystem could use functions of other PEIS devices in the environment, in order to compensate or complement for its own functions. Thus, the robots' capabilities for perception and action were facilitated by their direct communication and interconnection with the ambient intelligence, the subsystems and operational components distributed in the environment. For instance, the capability of a robotic entity to understand its position and manage objects in the environment depended on information received from the position sensors in other locations in the house (such as a camera on the ceiling), as well as the use of RFID tags placed on different objects, to provide information for their location and physical condition (shape, weight etc.). The project redefined the concept of the autonomous humanoid robot assistant that can perceive and take intelligent action; its distributed configuration could enhance functionality beyond the constraints of autonomous service robots.[88]

Such cooperative operations between robots and a sensor network can facilitate several possible functions, such as morning awakening of the user, monitoring and warning about expired food in the fridge, detecting and relocating obstacles in the robot's trajectory, control of incoming natural light and protection of a plant, management of artificial room illumination, and delivery of a book to handicapped users. Another activity that the researchers implemented and documented was the delivery of a drink to the inhabitant, while reading on his/her sofa. The system, after recognizing the presence and position of the inhabitant, would send a mobile robot to recognize his/her identity (with the help of a face recognition system and a camera mounted on the ceiling) and determine his/her favorite drink. Then a robotic table would approach to the fridge, to acquire the drink. Using an internal gripper, the camera and the activated fridge door, the table would acquire the drink and deliver it to the inhabitant (Figures 2.14 to 2.16). Although the operation seems very simple and slow,[89] the project operates as an assistive automated servant, providing simple but useful services in everyday domestic habitation.

The environment is capable of flexible adaptation to personalized changes, because users can easily connect new components to the system according to their needs. Since functionality is not centrally determined, but emerges from the cooperation of many simpler, specialized and distributed devices, users can flexibly modify it by adding new devices at any time. Once added, these devices publicize their presence and properties, so that other existing devices can detect

FIGURE 2.14 Living room with robot Astrid, PEIS Ecology, AASS Cognitive Robotics Systems laboratory, Orebro University, Sweden, 2008. Courtesy Alessandro Saffiotti.

FIGURE 2.15 Fridge with the transportation table, PEIS Ecology, AASS Cognitive Robotics Systems laboratory, Orebro University, Sweden, 2008. Courtesy Alessandro Saffiotti.

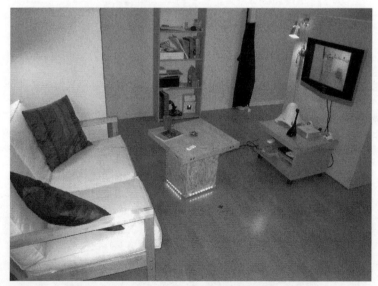

FIGURE 2.16 Living room with the transportation table, PEIS Ecology, AASS Cognitive Robotics Systems laboratory, Orebro University, Sweden, 2008. Courtesy Alessandro Saffiotti.

them, using a flexible and light open-source middleware.⁹⁰ The functions of the added components are combined with those of the existing devices. Thus, the overall functionality is changeable, provided that new components and robotic entities are embedded. As the researchers have explained, what PEIS Ecology has not yet been able to succeed is to recognize human activities and intentions, in order to adapt to them – for instance, if the system recognizes that the user intends to rest, it can turn off the vacuum cleaner or relocate it to a different room. In a future perspective, the system would be able to anticipate the changing physical capacities of inhabitants as they grow older and reconfigure its model of them, modifying its functionality accordingly.

As discussed above, in general there are significant developments in human activity recognition (albeit with some limitations), but not in the recognition of human intentions and desires, because of their indeterminacy and context-dependency. However, the capacity of the PEIS system to embody new added components and thus create new overall functionalities, means that it can possibly adapt to new and unexpected demands. This makes it significantly flexible, with many possible future research directions and potential. Users are thus capable of modifying and overriding the operational rules of the system, which is another promising direction in AmI research, as we shall see later. Before we come to that, we will address the social problems pertaining to the AmI vision.

Further Issues in Ambient Intelligence

Social Problems in AmI

Although intelligent environments are only experimental projects, ubiquitous computing is undeniably a fact of our daily life (think for instance of the extensive use of mobile phones, mobile internet, tablets and smart traffic systems in highways). Immaterial bits traveling through networks, messages, online exchanges, virtual agents, information services, all penetrate our social environment, with all sorts of electronic devices instantiating a technosocial assemblage of material artifacts, buildings and information. Several studies have examined the socio-political implications,⁹¹ as well as the social consequences of living and working in spaces that are augmented with information technologies, addressing, in particular, problems of social control, surveillance, and the violation of privacy.⁹²

Although AmI enhances the power of people to control their domestic environments, at the same time it may also take it away. By proactively responding to human needs, IEs can liberate people from the cognitive and physical burden of tedious, routine daily tasks. This, as Philip Brey has explained, gives users a sense of delegated control over the functionality of the system. On the other hand, the difficulties in activity and intention recognition, discussed previously, may make the system respond in incorrect ways, and therefore induce the psychological feeling of loss of control. This may also happen if collected user data in IEs are potentially used by third parties for interests other than those of the users, such as commercial purposes (which may happen if commercial firms own AmI applications).[93] Moreover, AmI technologies may disappoint their users, as their potential failure, instability and unreliability, an underestimated problem according to John Thackara, will inevitably raise the need for constant maintenance and technical support, thus enhancing further the problem of control.[94]

There are also several privacy and security issues. Some researchers raise concerns about the ubiquity of invisible sensor technologies in domestic environments that can keep records of users' past activities and behaviors, and wirelessly distribute information beyond the limits of the house. Therefore, and despite the fact that IEs are supposed to provide added security and safety (through smart surveillance systems able to detect intruders and alarm accordingly), users may inevitably experience privacy anxieties and the feeling of invasion into their personal affairs.[95] Similar issues can arise in intelligent workspaces, because digital technologies can operate as a sort of invisible "panopticon," which detects and controls employees' activities and behaviors.[96]

The above concerns are more prevalent in IEs that perform autonomous functions, namely systems that program themselves through learning and interaction with their environments to automate their operations (such as the iDorm). In this case there are intelligent agents that control and determine the overall behavior of the system and the way it manages incoming and outgoing information. But users cannot modify the functionality of those agents – they cannot intervene in the system software and learning algorithm.[97] However, many studies have shown that users of intelligent environments need to easily configure and control their functions,[98] and that the extended system autonomy can make users feel insecure about the reliability of the system, raising at the same time, and as discussed previously, privacy concerns. Users seem to feel the need for control; to be able to deactivate the overall system through a single switch, and limit system autonomy to very specific rooms in the house, such as the bathroom and the kitchen.[99]

This, however, may be the case even when smart surveillance systems are installed in domestic environments. These are different in perception from those used in the public domain, which are usually in accord with power structures and normalization systems. This is because domestic surveillance systems are installed voluntarily to provide security and privacy (which are acceptable concepts for domestic matters in western societies), and to benefit those under surveillance (particularly for fending off external threats and monitoring health issues). At the same time they operate as disciplinary instruments that facilitate the horizontal diffusion of power, in place of top-down control structures. Michele Rapoport explains that, as residents willingly and constantly supervise their private spaces, they are forced to take action in case some state of emergency occurs, thus participating actively in the experience of dwelling, along with a sense of power, deriving from the fact that being visible and monitored is a conscious choice, determined by those under surveillance.[100] But while public space surveillance detaches observing from observed subjects (who are rendered passive, objectified and suppressed), domestic surveillance rewrites traditional hierarchies of visibility, as users can be both agents and performers, observers and objects of observation, and as the fluidity of the focal point and the context of surveillance can transform its discursive meanings. In domestic surveillance, the site, users and technologies form a tripartite assemblage of components, engaging in fluid constant exchanges and transformations.[101]

Autonomous and End-User Driven IEs

The functions of autonomous IEs are determined by the capacities of intelligent agents, which are in control of the operational rules of the system. This is a prerequisite for buildings that should be able to make overall optimal decisions on their operation, when, for instance, human intervention is unimportant or insufficient. Autonomy has been considered to be useful and acceptable for safety and efficiency purposes – such as monitoring security and anticipating failure of supply networks, control of energy balance, and protection from external destabilizing factors, like earthquakes. On the other hand, the so-called *end-user driven* IEs are more functionally open, because control is given over to users, who can improvise and create functionalities according to their preferences.[102] In other words users are responsible for programming the functions of the environment. Even if they may not have programming skills or other technical knowledge, they can use the appropriate user interfaces to set the behavioral rules of the environment. Thus, choosing to install end-user driven IEs seems to be more reasonable in intelligent environment applications, because these systems encourage creative and personalized management of the environment, as well as "friendly" relationships between users and the system. Autonomous systems, on the other hand, benefit in that they operate in the background, saving the user from constantly having to engage and control its functions.[103]

End-user driven systems comprise user-friendly interfaces and interactions for simple applications in domestic environments. For instance, *PiP* (Pervasive interactive Programming) is an innovative method developed at Essex University that allows users to construct their own virtual domestic appliances (meta-appliances), combining parts of the functions of devices, in order to adjust their environment to respond to every particular need. The system is deconstructed in such a way as to allow users to choose, combine and group different functionalities of appliances (interacting either with the physical device directly, or with an interface that digitally represents the appliances), to create "virtual pseudo-appliances," the so-called *MAps* (Meta-Appliances Applications). Thus, the system proposes new complex functionalities of domestic appliances (e.g. when the phone rings the TV can show a pre-programmed image etc.). Based on its interactions with users and their desired behaviors, it creates a set of rules for each MAp, combining two different types of activities: an activation condition (if) and a consequence or resulting activity (then), i.e. propositions "if..., then...." These rules determine the synchronization of the activities of the functions of the devices, which provide the behavior of the virtual appliance.[104] Another important end-user driven IE, the *e-Gadgets*, will be discussed in Chapter 5.

Although there are certain benefits in end-user driven systems, some researchers believe that a more rational solution is the combination of autonomy and user control. This is because a combined model allows users to choose which parts of the system will be controlled by themselves, and which parts will be controlled by intelligent agents. In this case, the level of autonomy of the system is determined by users, thus optimizing the possibility for creative use (in the case of the end-user driven) and symbiosis (in the autonomous system).[105] Matthew Ball et al. have presented a conceptual model for a hybrid intelligent system, combining autonomous and end-user driven capacities, capable of learning from user activities and, at the same time, taking into account user-programmed rules. In this way, the model can anticipate users' desires, constructing behavioral rules, based either on user-programmed input (through a user-friendly interface), or autonomous learning, in order to change the operation of the environment accordingly. For its autonomous learning mechanism, the hybrid system uses a series of agents and an interpretative mechanism to track the users in the environment, in order to create a behavioral model that reflects their habitual activities. It can take into account information from multiple participants – simple user preferences, energy consumption quotas, dependencies between system mechanisms, health and safety limitations, as well as privacy issues. Depending on the degree of interest for each of these factors, the autonomy levels of the environment will be adjusted. If users are interested in some subsystem of the environment (e.g. the lighting system), then the autonomous learning mechanism will refrain from developing behavioral rules for this part of the system, and attribute its management to the user. The autonomous learning mechanism will be able to manage systems for which users have not expressed their interest.

To achieve this, Ball et al. proposed a theoretical sliding-scale switch, able to adjust the level of autonomy for each subsystem of the environment. If users move the switch somewhere in-between the two extreme ends – which represent either total autonomy or total user control – the system will be able to facilitate cooperation between users and intelligent agents, as well as to delegate management tasks to each other. Thus the system depends on the initiative and guidance of both the agents and the users. This is a mixed-initiative interaction, a hybrid autonomous and user-driven system capable of adjustable autonomy. This adjustment could be applied in every individual subsystem (lighting, security, heating, music choice etc.), to create a theoretical mixing desk with multiple switches, one for each subsystem, which will finally determine the amount of autonomy of the overall environment. Both users and intelligent agents cooperate through dialogical interaction, to find the optimal solution for programming the system operation, which is analogous to problem solving and conversation between two people.[106] In Chapter 4 we will further discuss the theoretical aspect of such a conversational approach to interaction.

Conclusions

Intelligent domestic environments can provide automated services to their potential inhabitants, especially people with special needs, by means of learning and proactive adaptation mechanisms. Such "serving" environments are functionally autonomous; they automate the management of domestic services, energy and optimization of functions.[107] On the other hand, IEs present difficulties in recognizing complex activities; although activity recognition is to a significant extent possible, in several cases, as we saw, AmI engineers must attach devices onto the inhabitants to get sufficient results. Consequently, the implementation of IEs presupposes technical demands that, for most people, would be a "burden," rather than a means to facilitate their daily lives and convenience. Since the recognition of intentions and desires, as well as the recognition of complex and simultaneous human activities, are problematic, the system cannot function efficiently in everyday environments.

At the same time, behind the potential of intelligent environments to autonomously recognize activities (albeit with some limitations) lies the danger of enforcing inhabitants to conform to the capacities and demands of the system. That is, there is the possibility that users, in order to facilitate the efficient functioning of the system, will direct their movements and activities to provide the appropriate stimuli that the system is able to recognize.

On the other hand, end-user driven IEs may be a more promising perspective for the recognition of human activities and the flexibility and adaptability of spaces, since the user provides the required information. In this case, feelings of insecurity that might arise in autonomous systems are also minimized. In Chapters 4 and 5 we will further address this aspect of AmI research, by looking into conversation theory in human–computer interaction, and theories from the social studies of technology. But before that, in the next chapter we will discuss the psychological aspects of augmented architecture that pertain to the human–machine boundary question.

Notes

1 See Cedric Price, *Cedric Price. Works II* (London: Architectural Association, 1984); Neil Spiller, *Visionary Architecture: Blueprints of the Modern Imagination* (London: Thames & Hudson, 2006); John Frazer, *An Evolutionary Architecture* (London: Architectural Association, 1995), 40–41.
2 Frazer, *An Evolutionary Architecture*, 41.
3 See the articles "World's First Intelligent Building," *RIBA Journal*, January, 1982; Deyan Sudjic, "Birth of the Intelligent Building," *Design*, January, 1981; "Cedric Price's Generator," *Building Design*, February 23, 1979.

4 Unlike Descartes who viewed the human mind as the only human organ that cannot be reduced to mechanism, Pascal and Leibniz opened the way for the implementation of mechanical computers, thus attributing a technological body to intelligence and rejecting Cartesian idealism. Leibniz in particular, in the 17th century, assimilated arithmetic reasoning with other forms of reasoning, such as ethical, legal, commercial, scientific and philosophical, on the assumption that each logical thought follows certain rules, albeit not arithmetic. He tried to analyze these rules to discover the language of thought and the operation of reasoning, that is, formal logic, in terms of arithmetic and alphabetical values. According to Leibniz, if all thoughts could be expressed arithmetically, it would be possible to mechanize any type of reasoning and thus produce a machine capable of thinking.
5 The Turing Test, proposed by Alan Turing in 1950, provided a satisfactory functional definition of intelligence, still effective today. The test was making use of the judgment of a human observer to determine whether he communicated with a machine or a human being. The observer would pose questions in written form and receive answers through a channel with two hidden correspondents – a human and a computer. If the examiner could not conclude where the written answers came from, the computer would pass the test – it would be considered to be intelligent. See Turing's seminal paper "Computing Machinery and Intelligence," *Mind* LIX, 236 (October 1950): 433–460.
6 See Warren S. McCulloch and Walter H. Pitts, "A Logical Calculus of the Ideas Immanent in Nervous Activity," in *Embodiments of Mind*, ed. Warren S. McCulloch (Cambridge, MA: MIT Press, 1965), 19–39. Originally published in *Bulletin of Mathematical Biophysics* 5, (1943): 115–133. McCulloch and Pitts showed that intelligence is embedded in the brain, and that certain neuronal networks calculate specific kinds of logical functions. Thus, they proposed a model of artificial neurons, capable of learning and computing any computable function.
7 See Stuart J. Russel and Peter Norvig, *Artificial Intelligence: a Modern Approach* (New Jersey: Pearson, 2003 [1995]), 3.
8 Allen Newel and Herbert Simon, "Computer Science as Empirical Inquiry: Symbols and Search," *The Tenth Turing Lecture, Communications of the Association for Computing Machinery* 19, 3 (1976): 113–126. Reprinted in John Haugeland, ed., *Mind Design: Philosophy, Psychology, Artificial Intelligence* (Cambridge, MA/London: MIT Press, 1981).
9 Johnston, *The Allure of Machinic Life*, 59–60.
10 See Margaret Boden, "Introduction," in *The Philosophy of Artificial Intelligence* (Oxford/New York: Oxford University Press, 1990), 7. Also Russel and Norvig, *Artificial Intelligence*, 12–14.
11 Marvin Minsky and Seymour Papert's 1969 book *Perceptrons: An Introduction to Computational Geometry* (Cambridge, MA: MIT Press) contributed significantly to making connectionism unpopular in scientific communities.
12 David Rumelhart and James McClelland's 1986 seminal book of two volumes *Parallel Distributed Processing* (Cambridge, MA: MIT Press) made PDP popular in the 1980s.
13 See for instance John McCarthy, "Ascribing Mental Qualities to Machines," in *Philosophical Perspectives in Artificial Intelligence*, ed. Martin Ringle (Brighton, Sussex: Harvester Press, 1979), 1–20; Marvin Minsky, *Semantic Information Processing* (Cambridge, MA: MIT Press, 1968).
14 John Searle, for instance, argued for the interdependence between intelligence and material substrate. See Searle, "Minds, Brains, and Programs," *Behavioural and Brain Sciences* 3, 3 (1980): 417–424.
15 Russel and Norvig, *Artificial Intelligence*, 2–5.
16 Knowledge-based or expert systems are computational systems able to manage large quantities of coded information, programmed to mimic the processes by which a human expert solves a problem in a delimited knowledge sector and a given database

(for instance medical diagnoses). These systems codify the best practices in a specific situation to advise human operators on the best action and help them confront, understand and use complex information sets efficiently.
17 Michelle Addington and Daniel Schodek, *Smart Materials and Technologies: For the Architecture and Design Professions* (Oxford: Architectural Press, 2005), 209.
18 See Kari Jormakka and Oliver Schurer, "The End of Architecture as we know it," in *The 2nd IET International Conference on Intelligent Environments (IE06)*, vol. 2, Athens, 5–6 July 2006 (London: IET publications, 2006), 109.
19 Maria Luisa Palumbo, *New Wombs: Electronic Bodies and Architectural Disorders* (Basel: Birkhäuser, 2000), 77–78.
20 See Dorothy Monekosso, Paolo Remagnino, and Kuno Yoshinori, eds., *Intelligent Environments: Methods, Algorithms and Applications* (London: Springer, 2008), 4.
21 With the theory of enaction, Varela attempted to overcome the deficiencies of the earlier stages in cognitive science (symbol processing and neural nets). He rejected the reliance of cognitive processes on representations of the world and restored the common sense unity between the body, its environment and cognitive processes. See Francisco Varela, *Invitation aux sciences cognitives* (Paris: Éditions du Seuil, 1996).
22 See for instance George Lakoff and Mark Johnson, *Philosophy in the Flesh: The Embodied Mind and its Challenge to Western Thought* (New York: Basic Books, 1999). Lakoff and Johnson argued that the body forms our cognitive structures, and that thinking is based on metaphors that derive from our environmental experiences and embodied interactions with the world.
23 Andy Clark, *Being There: Putting Brain, Body, and World Together Again* (Cambridge, MA: MIT Press, 1997).
24 See for instance Searle, "Minds, Brains, and Programs."
25 See Rolf Pfeifer and Josh C. Bongard, *How the Body Shapes the Way We Think: A New View of Intelligence* (Cambridge, MA: MIT Press, 2006); Rodney Brooks, *Cambrian Intelligence: The early History of the new AI* (Cambridge, MA: MIT Press, 1999).
26 Apart from Varella's *Invitation aux sciences cognitives,* see Francisco Varela, Evan Thompson, and Eleanor Rosch, *The Embodied Mind: Cognitive Science and Human Experience* (Cambridge, MA: MIT Press, 1992). Extending autopoietic theory, Varela, Thompson and Rosch argued that cognition can be understood only in relation to the body and the natural environment with which it interacts, thus emphasizing the interconnections between the human nervous system and the environment.
27 For a general account of this view see Monica Cowart, "Embodied Cognition," *The Internet Encyclopedia of Philosophy*, accessed April 5, 2013, http://www.iep.utm.edu/e/embodcog.htm. Of course, the idea of an embodied mind has been a significant feature of phenomenology and the philosophy of embodiment. See Martin Heidegger, *Being and Time* (New York: Harper and Row, 1962 [1927]). See also Hubert L. Dreyfus, *Being-in-the-World: A Commentary on Heidegger's Being and Time, Division I* (Cambridge, MA: MIT Press, 1991).
28 Kevin Kelly, *Out of Control: The Rise of Neo-biological Civilization* (Reading, MA: Addison-Wesley, 1994), 24.
29 See Takashi Gomi, "Aspects of non-Cartesian Robotics," *Journal of Artificial Life and Robotics* 1, 2 (1997): 95–103.
30 Rodney Brooks, "A Robust Layered Control System for a Mobile Robot," in *Cambrian Intelligence: The early History of the new AI* (Cambridge, MA: MIT Press, 1999), 3–26. Also see Rodney Brooks, "Intelligence without Reason," in *Cambrian Intelligence*, 133–159. In subsumption architecture every behavior represents a single action, performed according to a hierarchy of priority levels, and determined by the associations between sensor input and actuator output. This hierarchy selects a higher-level behavior, suppressing those of lower importance. Such a prioritized succession of behaviors, for example, can be represented by the following activities (starting with the higher level in the hierarchy): "eat-food," "explore-the-world," "wander," "avoid-obstacle." Each of these modes, having access to the sensor data,

activates the actuators. The higher levels suppress low-order activities according to sensor input, taking control of their respective actuator output. Thus, lower levels can operate as reflective mechanisms, so that the higher levels can achieve the overall target (for instance, the level "eat-food" being higher in the hierarchy than the level "avoid-obstacle," will subsume the latter so that the robot will continue moving to obtain the initial objective).

31 Brooks, "Intelligence without Reason," 149.
32 On the introduction of this idea in architectural environments see Stephen Jones, "Intelligent Environments: Organisms or Objects?" *Convergence: The International Journal of Research into New Media and Technologies* 7, 2 (2001): 25–33. Also Ted Krueger, "Eliminate the Interface," in *Gamesetandmatch Conference Proceedings*, eds. J.C. Hubers, Misja van Veen, Chris Kievid, and R. Siemerink (Publikatiebureau Bouwkunde, Faculty of Architecture DUT, 2004), accessed July 21, 2014, http://www.bk.tudelft.nl/fileadmin/Faculteit/BK/Over_de_faculteit/Afdelingen/Hyperbody/Game_Set_and_Match/GameSetandMatch_1/doc/gsm_I.pdf.
33 For instance see Panagiotis Stamatis, Ioannis Zaharakis, and Achilles Kameas, "Exploiting Ambient Information into Reactive Agent Architectures," in *The 2nd IET International Conference on Intelligent Environments (IE06)*, vol. 1, Athens, 5–6 July 2006 (London: IET publications, 2006), 71–78.
34 See Oungrinis, Δομική Μορφολογία, 359–360. The responsive and kinetic system that Oungrinis speculated on, engaged multiple parallel subsumption agents, with simple behaviors, combined with a central agent that would be able to reprogram the agents and supervise the overall outcome. This agent would also optimize itself, by collecting and analyzing data to avoid "collisions." On a lower functional level, the building would attempt to fulfil the desires of its inhabitants, both in terms of physical changes (such as change of spatial boundaries), as well as perceived changes (such as changes of color and lighting). In an email message to the author (February 13, 2010), Oungrinis explained his idea on the basis of biological models and the adaptive functions of organisms. He suggested that buildings should be able to adapt to structural balance changes, by means of an energy absorbing, flexible frame, and optimize energy consumption, by means of surface panels able to filter the incoming energy. Also the internal space should be separated from the building frame to attain flexibility, just as internal organs float within the body of cells to protect themselves from intrusions. Oungrinis also compared the circulatory and neurological systems of organisms to the circulatory and intelligent communication network in buildings, for the transfer of the necessary elements and information. The latter comparison recalls Addington and Schodek's proposal for the design of intelligent environments, modeled on the neurological processes of the human body. This kind of environment would embody sensors and actuators with singular or multiple behaviors, able to interact and mutually affect their respective actions and reactions. See Addington and Schodek, *Smart Materials and Technologies*, 215–216.
35 Of course the experiments of the new AI and multi-agent systems, seem to implement the ambitions of early cyberneticists, like Ross Ashby and John von Neumann who, back in the 1950s, experimented either practically or theoretically with self-organizing machines and self-reproducing automata.
36 On ethology see also David McFarland, *Animal Behaviour: Psychobiology, Ethology, and Evolution* (New York: Longman, 1998).
37 On the behavior and function of ant colonies see Jean Louis Deneubourg et al., "The Self-Organizing Exploratory Pattern of the Argentine Ant," *Journal of Insect Behavior* 3, 2 (1990): 159–168.
38 *Boids* (short for "birdoids") is a computer program designed by Craig Reynolds in 1987. Artificial digital units develop flocking behavior similar to that of birds, although nothing in the program models the overall behavior of the flock. On the contrary, the behavior of every single organism is programmed to follow three simple laws that pertain to distance, perception of speed and direction of movement

of the neighbors. See the *Boids* webpage at http://www.red3d.com/cwr/boids. Also see Eric Bonabeau and Guy Théraulaz's report "Swarm Smarts" (*Scientific American* 282, March 2000, 72–79) for the successes in using swarms as models for the design of problem-solving software.
39 Eric Bonabeau, Marco Dorigo, and Guy Théraulaz, *Swarm Intelligence: From Natural to Artificial Systems* (New York: Oxford University Press, 1999). Also see Marco Dorigo and Mauro Birattari, "Swarm Intelligence," *Scholarpedia* 2, 9 (2007): 1462, doi:10.4249/scholarpedia.1462.
40 Eugene Thacker, "Networks, Swarms, Multitudes: part one," ctheory, 2004, accessed February 2, 2013, http://www.ctheory.net/articles.aspx?id=422; Eugene Thacker, "Networks, Swarms, Multitudes: part two," ctheory, 2004, accessed February 2, 2013, http://www.ctheory.net/articles.aspx?id=423.
41 Kas Oosterhuis, "A New Kind of Building," in *Disappearing Architecture: From Real to Virtual to Quantum*, eds. Georg Flachbart and Peter Weibel (Basel: Birkhäuser, 2005), 95.
42 Ibid., 106–107.
43 Ibid., 96.
44 Chris Kievid and Kas Oosterhuis, "Muscle NSA: A Basis for a True Paradigm Shift in Architecture," in *Hyperbody. First Decade of Interactive Architecture*, eds. Kas Oosterhuis et al. (Heijningen: Jap Sam Books, 2012), 448.
45 Ibid., 449.
46 Oosterhuis, "A New Kind of Building," 109.
47 Kari Jormakka, *Flying Dutchmen: Motion in Architecture* (Basel: Birkhäuser, 2002), 21.
48 Oosterhuis, "A New Kind of Building," 107.
49 See the paper that Carranza co-authored with Paul Coates titled "Swarm modelling – The use of Swarm Intelligence to generate architectural form," presented at the 3rd International Conference in Generative Art in Milano, 2000. See also his website at www.armyofclerks.net. In an email message to the author (December 15, 2014), Pablo Miranda Carranza explained that the drawing was part of a diploma project at Unit 6, UEL (University of East London) School of Architecture under the supervision of Paul Coates. It shows the information left by agents on a three-dimensional array, equivalent to the pheromone trails left by ants in Ant Colony Optimization (ACO) algorithms and other computational ant simulations. The agents have an algorithm that detects collisions in order to be able to navigate a 3D model of the site (the Bow Creek area in the London East End), and each of them writes their heading at the current cell in the lattice (three-dimensional array) they are at. They also read the direction left by the previously passing agents through the cell they are at and add it to their current heading. The trails are programmed to "evaporate" with time, as in most sematectonic or stigmergic systems. This function has been compared with the capacity of the system to "forget." Also, a diffusion algorithm smoothens out the traces left by the agents in the lattice, in order to filter out noise in the behavior of the agents.
50 See the related websites at www.swarm-bots.org and mrsl.rice.edu.
51 See Marco Dorigo et al., "SWARM-BOT: Design and Implementation of Colonies of Self-assembling Robots," in *Computational Intelligence: Principles and Practice*, eds. Gary Y. Yen and David B. Fogel (New York: IEEE Computational Intelligence Society, 2006), 103–136.
52 For further discussion of the project see Socrates Yiannoudes, "An Application of Swarm Robotics in Architectural Design," in *Intelligent Environments 2009: Proceedings of the 5th International Conference on Intelligent Environments*, eds., Victor Callaghan, Achilleas Kameas, Angélica Reyes, Dolors Royo, and Michael Weber (Amsterdam: IOS Press, 2009), 362–370.
53 The installation was exhibited at the FRAC Centre in Orléans, France from December 1 to 4, 2011, and was part of the Aerial Construction research project, a collaboration between the Institute for Dynamic Systems and Control (http://www.

idsc.ethz.ch/Research_DAndrea/Aerial_Construction), and the Chair of Architecture and Digital Fabrication at the Department of Architecture ETH Zurich (http://www.gramaziokohler.arch.ethz.ch/web/e/forschung/240.html).

54 For a description of the installation see Federico Augugliaro, Sergei Lupashin, Michael Hamer, Cason Male, Markus Hehn, Mark W. Mueller, Jan Willmann, Fabio Gramazio, Matthias Kohler, and Raffaello D'Andrea, "The Flight Assembled Architecture Installation: Cooperative construction with flying machines," *IEEE Control Systems* 34, 4 (2014): 46–64.
55 For definitions on the vision of AmI see the website of the Information Society and Technologies Advisory Group (ISTAG) at: http://cordis.europa.eu/fp7/ict/istag/home_en.html. In particular see its 2001 and 2003 reports: "Scenarios for Ambient Intelligence in 2010," European Commission CORDIS, accessed July 8, 2014, ftp://ftp.cordis.europa.eu/pub/ist/docs/istagscenarios2010.pdf; "Ambient Intelligence: from Vision to Reality," European Commission CORDIS, accessed July 8, 2014, ftp://ftp.cordis.europa.eu/pub/ist/docs/istag-ist2003_consolidated_report.pdf.
56 See Mark Weiser, "The Computer of the 21st Century," *Scientific American* 265, 3 (1991): 66–75.
57 Jari Ahola, "Ambient Intelligence," Special Theme: Ambient Intelligence – *ERCIM News* 47, (October 2001): 8, http://www.ercim.org/publication/Ercim_News/enw47/intro.html.
58 See the Intelligent Environments website at: http://www.intenv.org.
59 Diane J. Cook and Sajal K. Das, "How Smart are our Environments? An updated look at the state of the art," *Pervasive and Mobile Computing* 3, 2 (2007): 53–73.
60 Ibid., 53.
61 Addington and Schodek, *Smart Materials and Technologies*, 19.
62 Ibid., 204–208.
63 Ahola, "Ambient Intelligence," 8.
64 Addington and Schodek, *Smart Materials and Technologies*, 211.
65 See Ted Krueger, "Editorial," *Convergence: The International Journal of Research into New Media and Technologies* 7, 2 (2001). Also see Ranulph Glanville, "An Intelligent Architecture," *Convergence: The International Journal of Research into New Media and Technologies* 7, 2 (2001): 17.
66 In this category, we could also place intelligent structures that can control and optimize the efficiency and behavior of their load-bearing system against earthquakes and other dynamic strains. Michael Fox mentions that kinetic systems are important for this sort of structural flexibility and behavior of buildings. Citing engineer Guy Nordenson, he argues that if an intelligent kinetic structural system, in a building like a skyscraper, could constantly react actively and locally to external natural forces, it could minimize material and load-bearing mass. See Fox, "Beyond Kinetic."
67 See the project webpage at grimshaw-architects.com/project/rolls-royce-manufacturing-plant-headquarters.
68 See the related websites at www.intel.com/research/exploratory; www.cs.colorado.edu/~mozer/nnh; web.mit.edu/cron/group/house_n/placelab.html; ailab.wsu.edu/mavhome.
69 Addington and Schodek, *Smart Materials and Technologies*, 208. See also Fox, "Beyond Kinetic" and his description of the Moderating Skylights project, a kinetic louver roof with selective response.
70 See for instance C. C. Sullivan, "Robo Buildings: Pursuing the Interactive Envelope," *Architectural Record* 194, 4 (2006): 156.
71 John Thackara, *In the Bubble: Designing in a Complex World* (Cambridge, MA/London: MIT Press, 2005), 204.
72 The Smart-Its platform consists of micro-devices that can be attached on any object in space. It comprises different sensors, wireless communication and micro-processors that can decide and evaluate particular local situations. See Michael Beigl,

"Ubiquitous Computing: Computation Embedded in the World," in *Disappearing Architecture: From Real to Virtual to Quantum*, eds. Georg Flachbart and Peter Weibel (Basel: Birkhäuser, 2005), 57.
73 Achilles Kameas, interviewed by the author, September 16, 2009.
74 See the website at cswww.essex.ac.uk/iieg/idorm.htm.
75 See Hani Hagras, Victor Callaghan, Martin Colley, Graham Clarke, Anthony Pounds-Cornish, and Hakan Duman, "Creating an Ambient-Intelligence Environment using Embedded Agents," *IEEE Intelligent Systems* 19, 6 (2004): 12–20.
76 Faiyaz Doctor, Hani Hagras, Victor Callaghan and Antonio Lopez, "An Adaptive Fuzzy Learning Mechanism for Intelligent Agents in Ubiquitous Computing Environments," *Proceedings of the 2004 World Automation Conference*, vol. 16, Sevilla, Spain, 2004 (IEEE publications, 2004), 108–109.
77 Victor Callaghan et al., "Inhabited Intelligent Environments," *BT Technology Journal* 22, 3 (2004): 242. The procedure involved the "learning inertia" routine (a parameter that the user could adjust). The agent would not adapt its acquired rules until the changes that the user conducted on an actuator value set reappeared several times. This prevented the agent from adapting its rules right after a single user activity, because this would not reflect a profound change in his/her standard behaviors.
78 The way the agent operated was evaluated according to how well it adjusted the environment to the preferences of the user. The agent derived this information from observed user activities, recorded during the initial three days of observation and the second two-day phase. As the results showed, user intervention, which was initially high, was diminishing through time, and was finally stabilized early afternoon of the second day. The agent started with the 186 rules that it learned in 408 training instances, which were gathered during the first three days of monitoring the user. During the next two days another 120 new rules were added, and the agent could learn and discreetly adapt to most of the user preferences for different environmental conditions and activities, such as lying on the bed, listening to music or sitting on the desk to work on the computer. See Doctor et al., "An Adaptive Fuzzy Learning Mechanism," 109–110.
79 There have been several attempts to resolve this problem, such as the design of algorithms that can probabilistically recognize simultaneous activities in real life. See Geetika Singla and Diane J. Cook, "Interleaved Activity Recognition for Smart Home residents," in *Intelligent Environments 2009*, 145–152. There is also research for the development of systems that can learn patterns of connected human activities, in order to discover frequent sets of simultaneous activities. See Asier Aztiria et al., "Discovering Frequent Sets of Actions in Intelligent Environments," in *Intelligent Environments 2009*, 153–160.
80 Cook and Das, "How Smart are our Environments?" 67.
81 Ibid., 58. Also see Diane Cook, Kyle D. Feuz, and Narayanan C. Krishnan, "Transfer Learning for Activity Recognition: A Survey," *Knowledge and Information Systems* 36, 3 (2013): 537–556; Diane Cook, Narayanan Krishnan, and Parisa Rashidi, "Activity Discovery and Activity Recognition: A New Partnership," *IEEE Transactions on Cybernetics* 43, 3 (2013): 820–828.
82 See the website at web.mit.edu/cron/group/house_n/intro.html.
83 See Kent Larson and Richard Topping, "PlaceLab: A House_n + TIAX Initiative," House_n, accessed May 24, 2013, http://architecture.mit.edu/house_n/documents/PlaceLab.pdf.
84 Stephen Intille et al., "The PlaceLab: a Live-In Laboratory for Pervasive Computing Research (Video)," *Proceedings of Pervasive 2005 Video Program*, Munich, Germany, May 8–13, 2005, accessed January 27, 2009, http://web.media.mit.edu/~intille/papers-files/PervasiveIntilleETAL05.pdf.
85 Beth Logan et al., "A Long-Term Evaluation of Sensing Modalities for Activity Recognition," in *Proceedings of the International Conference on Ubiquitous Computing*, LNCS 4717 (Berlin/Heidelberg: Springer-Verlag, 2007), 484–485.

86 Ibid., 486–487.
87 Ibid., 491–499.
88 Alessandro Saffiotti et al., "The PEIS-Ecology Project: Vision and Results," In *Proceedings of the IEEE/RSJ International Conference on Intelligent Robots and Systems, IROS 2008*, Nice, France, September 2008 (IEEE publications, 2008), 2329.
89 See the video at aas.oru.se/pub/asaffio/movies/peis-ipw08.flv.
90 Middleware is a type of software that can mediate between a software application and a network, to enable interaction between dissimilar applications in the framework of a heterogeneous computational platform.
91 Contemporary IEs are subject to social critique because of the implied ideological intentions that promote them and that they reproduce. Lynn Spigel, for instance, explains that smart homes promise the preservation and expansion of traditional middle-class convenience, by means of division of labor, and distinction between those who serve and those who are served. Spigel argues that, although smart homes are marketed and promoted with liberating connotations, they reproduce the distinction between human and technology, and they enhance the division of labor between the two sexes. At the same time they use internet technology to monitor and control their boundaries. See Lynn Spigel, "Designing the Smart House: Posthuman Domesticity and Conspicuous Production," *European Journal of Cultural Studies* 8, 4 (2005): 403–426. Similar kinds of critiques come from scholars who have argued that the concept of intelligent machines that serve a class of humans, reproduces the problematic relations between those that serve and those that are served. Alexandra Chasin, for instance, tracing the relations between forms of mechanical (re)production (from the mechanical, to the electrical and the electronic), forms of labor (from industrial to service) and the differences between humans and machines, argued that electronic technology stabilizes, enhances and participates in the idea that a class of servicing beings (either human or technological) is appropriate and even necessary. Although smart machines point to the expectation that people will be liberated from work, their increasing distribution and use requires multiple service relations. And this leads to the increase of human labor (for technical support and cheap manufacturing), with inhuman working conditions. See Alexandra Chasin, "Class and its close relations: Identities among Women, Servants, and Machines," in *Posthuman Bodies*, eds. Judith Halberstram and Ira Livingstone (Bloomington: Indiana University Press, 1995), 93–175. Following Chasin, Lucy Suchman explained that, as the "smart" machine presents itself as a device that saves labor, at the same time it erases the particular socio-material infrastructures (bank clerks, software programmers, and Third World workers) that are necessary for its production and operation. See Lucy Suchman, *Human–machine Reconfigurations: Plans and Situated Actions* (Cambridge: Cambridge University Press, 2007), 225.
92 See Victor Callaghan, Graham Clarke, and Jeannette Chin, "Some Socio-Technical Aspects of Intelligent Buildings and Pervasive Computing Research," *Intelligent Buildings International Journal* 1, 1 (2009): 56–74.
93 Philip Brey, "Freedom and Privacy in Ambient Intelligence," *Ethics and Information Technology* 7, 3 (2006): 157–164.
94 John Thackara, the director of *Doors of Perception*, wrote that the vision of AmI is based on the unjustified assumption that all people and systems will constantly behave optimally, and therefore their function will be anticipated. But, because of the expanding speed of AmI, its enthusiastic researchers cannot see that the vast amount of networked microprocessors, sensors and smart materials constituting their applications, may be subject to constant failures and functional problems. In reality no technology has ever – and will ever – work efficiently, especially networked and complex technologies, which must constantly resolve compatibility and cooperation issues. See Thackara, *In the Bubble*, 205–206, 209.
95 Brey, "Freedom and Privacy in Ambient Intelligence," 164–166.

96 Lucy Bullivant, "Intelligent Workspaces: Crossing the Thresholds," *Architectural Design* (*4dspace*: *Interactive Architecture*) 75, 1 (2005), 39.
97 In the iDorm, as we saw, users can intervene when they are not satisfied with the system behavior, and change the functions of the devices manually. Then the system records these changes and adapts accordingly.
98 Jeannette Chin, Vic Callaghan, and Graham Clarke, "A Programming-by-Example Approach to Customising Digital Homes," in *Proceedings of the 4th IET International Conference on Intelligent Environments*, Seattle WA, 21–22 July 2008 (London: IET publications, 2008), 1–8.
99 Veikko Ikonen, "Sweet Intelligent Home – User Evaluations of MIMOSA Housing Usage Scenarios," in *The 2nd IET International Conference on Intelligent Environments (IE06)*, vol. 2, Athens, 5–6 July 2006 (London: IET publications, 2006), 13–22. Also Cook and Das make clear that there is still a lot of research that scientists should do, in order to make sure that the collected data, in intelligent environments, do not endanger the privacy and well-being of their inhabitants. See Cook and Das, "How Smart are our Environments," 68.
100 Michele Rapoport, "The Home Under Surveillance: A Tripartite Assemblage," *Surveillance & Society* 10, 3–4 (2012): 328–329, http://library.queensu.ca/ojs/index.php/surveillance-and-society/article/view/tripartite/tripartite.
101 Ibid., 330–331.
102 They do not have to conform to the particular functionalities of automated technological environments like Microsoft's "smart home" at Seattle, which the author had the chance to visit. This house could automate and anticipate needs, but users could not modify its specific functions. See also https://www.youtube.com/watch?v=9V_0xDUg0h0.
103 Matthew Ball et al., "Exploring Adjustable Autonomy and Addressing User Concerns in Intelligent Environments," in *Intelligent Environments 2009*, 430–431. Also see Matthew Ball and Vic Callaghan, "Explorations of Autonomy: An Investigation of Adjustable Autonomy in Intelligent Environments," in *2012 8th International Conference on Intelligent Environments (IE)*, Guanajuato, Mexico, 26–29 June 2012 (IEEE publications, 2012), 114–121.
104 Chin, Callaghan, and Clarke, "A Programming-by-Example Approach to Customizing Digital Homes."
105 Ball et al., "Exploring Adjustable Autonomy," 431–432. Ball et al. also argue that human use, and the extent of intimacy between users and both autonomous and end-user driven systems, may present certain problems. For instance, systems may intimidate users and result either in sabotage (in the autonomous system), or in abuse (in the end-user driven system).
106 Ibid., 432–435.
107 In this case, the system should be able to recognize basic activities (sleeping, working on the computer, eating), and monitor abnormal activities or deviations from habitual ones, in order to provide necessary and proper assistance, warnings and consultation on time. See Fernando Rivera-Illingworth, Victor Callaghan, and Hani Hangras, "A Neural Network Agent Based Approach to Activity Detection in AmI Environments," *IEE Seminar on Intelligent Building Environments* 2, 11059 (2005): 92–99.

3
PSYCHO-SOCIAL ASPECTS

Environments that employ sensors and actuators can perceive their context and react to human activities, thus demonstrating animate behavior. In this chapter we will address the psychological implications of augmented architecture, by exploring the perception of animacy in inanimate entities and artificial life objects. We will argue that the drive to design and construct animate and interactive structures and spaces is associated to the idea of the boundary object, which is discussed by sociologists of technology in relation to "marginal" computational objects. It is also part of the history and theory of artificial life practices since the 18th century, and can be further associated to the theoretical criticism of modernity's object–subject or nature–culture divide and human–machine discontinuities. We conclude the chapter by questioning the psychological implications that pertain to computationally augmented architecture.

The Psychological Aspect

E-motive House

Researcher Nimish Biloria, reviewing some of the experimental projects of the Hyperbody Research Group, concludes that "The successful accomplishment of these projects… stresses the need for further envisioning of *e*-motive architectural beings which understand and respond to their occupants."[1] The experimental projects he talks about (like the Muscle projects) can physically act and react to the activities of people, by means of sensors and actuators. In this sense, they are no longer perceived as inert inanimate things, but rather as entities that demonstrate some level of intelligence, challenging the boundaries between the animate and the inanimate. For instance, the *E-motive house* is a speculative design proposal for a transformable house, by Kas Oosterhuis and his ONL team, that can be programmed to respond in real time to the actions, needs and desires of its inhabitants, providing a space for work, food or sleep.[2] It is depicted as a modular shell, where walls, floor and ceiling can transform according to changing functional requirements; part of the floor can lift to become a working bench, a seating room or a dinner table. According to Oosterhuis, the structure of the house will comprise "soft" and "hard" construction parts. The former will be long-shaped inflatable chambers, placed between a number of massive wooden beams (connected by pneumatic actuators) that make up the hard structure, which can extend or shrink, assisted by cooperating hydraulic cylinders, to change the overall shape of the house. Like the Muscle NSA, the e-motive house system will run a Virtools software script, with rules and variable parameters that control the real-time interactions between visitors, actuators and external influences.

Although the technical description of the house seems vague and speculative, what is interesting here is the way Oosterhuis describes the behavior of the house. He implies that the house is a living organism, a "being" with changing emotional states, able to cooperate, learn, communicate and participate in social interactions with occupants. Oosterhuis mentions the possible objects of discussion between its residents:

> What mood is your house in today? Isn't it feeling well? Why is your house behaving so strangely lately? Perhaps it needs to see a doctor? Did you care enough for your house? Is your house boring you? Are you neglecting your house? Is your house suggesting that you might be boring in the way you perceive it? These would be the sort of social conversation topics between the inhabitants of e-motive houses.[3]

Because of the complex interactions between all the factors that affect its performance, the behavior of the house will be unanticipated and seemingly unpredictable, giving the impression of an emotional and social entity. By interacting with people and using its in-built intelligence, it will gradually develop a character by expressing a predefined series of psychological states. Oosterhuis emphasizes this behavioral aspect of the Hyperbody projects as well; he suggests that they reproduce and resemble the *behavior*, not the form of biological organisms: "We never try to copy superficially the appearance of a biological species. Rather we try to invent new species which by its complexity and complex behaviour may eventually start to familiarize with living objects as we already know."[4] This psychological dimension is also evident in the descriptions and representations of the other Muscle projects. For instance, the YouTube webpage video of the Hyperbody projects shows an implicit intention to present them as living entities, by emphasizing the constancy of their fluid deformations.[5] Moreover, Hyperbody's researchers MarkDavid Hosale and Chris Kievid, discussing the *InteractiveWall* (see also Chapter 4 for further review), point out that "It is through projects such as the InteractiveWall prototype that architecture will come alive."[6] Similarly, in a website developed for the commercial promotion of the *Hyposurface*, a programmable, transformable, responsive surface, a quote by Jody Levy (O2 Creative Solutions, Detroit) reads: "The HypoSurface™ itself is like an organism; responsive, stunning and organic in nature. It is captivating and full of endless possibilities."[7] Rem Koolhaas also made similar comments; he wrote that, while architecture was historically inert – its components deaf and mute – the pervasive ubiquity of the digital in buildings and cities has turned their constituting elements into active – and sometimes even untrustworthy – pieces that can listen, think, talk back, collect information and act accordingly: "While we still think of buildings as neutral spaces, our houses are assuming a degree of consciousness – the 'intelligent' building is a euphemism for an agent of intelligence. Soon, your house could betray you."[8]

To further study these psychological aspects we will now look at the body of experimental work that deals with the perception of animacy and anthropomorphic ascriptions on inanimate objects. As we shall see, a significant amount of experimental research findings in common sense psychology, experimental psychology (psychophysics), and neurophysiology have shown that seemingly autonomous self-generated motion, reactivity, as well as a number of other factors, contribute to the perception of objects as alive, animate entities.

Anthropomorphism and the Perception of Animacy

Anthropomorphism (from the Greek word "Anthropos" for man, and "morphe" for form/structure) is the psychological tendency to attribute human characteristics, cognitive or emotional states to inanimate entities (objects, machines, animals etc.). It deals with unfamiliar and intangible situations, by rationalizing the behavior of the other in a given social environment.[9] Scientists have commonly regarded anthropomorphism as hindrance to science and unworthy of rigorous study in psychology, because it complicates perceptual distinctions between the physical and the psychological, or the human and the non-human. It is, however, a trait of human evolutionary development – an "irresistible" and almost "inevitable," innate human tendency. People tend to ascribe anthropomorphic properties when they cannot explain the observed behavior and function of systems in mechanical and physical terms – in other words, when the system's internal structure and operations are *opaque*. The need to make sense of opaque, highly complicated and thus unintelligible objects, such as computational systems, may "generate an apparently nonadaptive social explanation (i.e., some personality characteristics or emotions) for observed machine behavior,"[10] as Linnda Caporael has explained, or even mental state attributions. Discussing anthropomorphism in the context of common sense psychology, Stuart Watt has argued that a hidden inner structure, or a highly complicated structure in a system, may invite mental state attributions.[11]

In the light of these ideas, we can discuss anthropomorphism in relation to the project *ADA – The Intelligent Space*. This was a computationally augmented space, developed for the Swiss EXPO.02 in 2002.[12] The designers of ADA wanted to create a system able to sense, locate, recognize and communicate with its visitors, as well as express a series of internal emotional states, through visual and auditory effector devices and events. Although the project will be further reviewed in Chapter 4, here we will address its animate and anthropomorphic aspects.

The designers described ADA as an "artificial creature."[13] They considered their work as "an exploration in the creation of living architecture,"[14] and explained that its visitors may perceive it as alive. Kynan Eng, a member of the development team, argued that this capacity was due to the density and opaqueness of its technological apparatus, which may have led to the perception of the project as animate. As he further explained, when the components of an intelligent building reach a critical degree of density of space and interconnectedness, the overall function may become opaque, as users and non-experts are no longer able to control it, although they can perform actions in its subsystems and observe consequences and interrelations: "From the viewpoint of the user, the building becomes more than the sum of its shell and subsystems; it appears to be alive."[15] He addressed the concept of opaqueness, as discussed previously, linking the lack of knowledge of the internal system structure to anthropomorphic attributions and perceptual animacy. In particular, the project team evaluated the attribution of life and intelligence to ADA by handing questionnaires to the visitors, in order to trace their perceptual experiences and psychological reactions. One of the questions was set to determine the degree to which visitors perceived the environment as "a kind of creature." In a scale from 1 (totally negative attribution) to 10 (totally positive attribution), the mean value was 4.61, which meant that the sense of life did occur among the visitors, but not to a significant extent.[16] As the researchers explained, through ADA's performance, they wanted to explore the limits of human interaction with an intelligent entity, and whether this interaction is possible with a system that does not look or behave like humans. The results verified what we saw before, that anthropomorphism, as a psychological tendency, does not necessarily need human-like forms to occur, but rather specific stimuli that induce irresistible mental or emotional state attributions.

Psychophysical findings in perceptual animacy support the aforementioned link between system opaqueness and anthropomorphic attributions. These findings have repeatedly demonstrated that knowledge of the causes of the motion of objects is critical for the attribution of animacy. This is explained by the fact that our visual system automatically detects events that appear to demonstrate self-initiated (autonomous) motion, i.e. increasing kinetic energy, without any apparent influence from an external energy source.[17] There is a significant body of experimental research on the perception of animacy in psychophysics – Fritz Heider and Mary-Ann Simmel's experiment being the

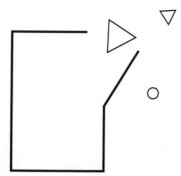

FIGURE 3.1 Display of moving geometric shapes

most well known[18] – which have employed displays of moving simple geometric shapes, in order to study our innate tendency to attribute animacy – personality features, intentions, mental and emotional states on inanimate entities, on the basis of their causal space–time relationship configurations (Figure 3.1).[19] Researchers have shown that short and simple motion paths that contain a simultaneous change in speed and direction, in the absence of any visible features that might explain this velocity change, are the minimal characteristics that can induce an impression of volitional control over the motion of the object. This implies the capacity for intentional goal-directed behavior, which is normally exhibited by living entities.[20] Although some researchers have questioned the significance of these findings, as well as their validity in real-life settings beyond the controlled environment of the lab,[21] their applications in robotics and system design are useful for our investigation.

In the fields of robotics and interface design, there are arguments for and against the use of human-like traits as metaphors in the design of interfaces or robots. On the one hand, this approach may establish false expectations and misunderstandings about an inanimate entity's behavior, but on the other it may be an effective design method to facilitate certain social modes of interaction.[22] Researchers that align with this latter view have argued that designers and engineers can make use of anthropomorphism in robotics by employing those physical and behavioral properties – stereotypical communication cues like nodding or shaking of the head – in order to create "sociable" robots, capable of communication and interaction with humans.[23] In line with the evidence from psychophysics discussed previously, as well as with findings from the neurophysiology of artistic experience,[24] researchers in robotics have shown that designers do not have to imitate the human structure, form or motion; slight unpredictable movement and orientation shifts of the robot, as well as caricaturized, simplified or mechanistic-looking forms, can induce compelling perceptions of robots as socially interactive, by surprising people or drawing their attention.[25]

We can now easily shift from robotics to kinetic actuator-driven structures, and attempt to determine their perceptual impact, on the basis of the previous findings. Of course, most examples of kinetic architecture, such as mobile,

FIGURE 3.2 *Animaris Percipiere*, Theo Jansen, Strandbeest. © Theo Jansen.

deployable or convertible structures, and most conventional building elements – such as windows and blinds that open and close momentarily – are not perceptually animate. Animacy could occur in structures that use robotic systems and actuators, such as Hyperbody's Muscle projects. Animacy would depend on whether their kinetic mechanism is visually exposed, and on the determination of its energy source, namely whether it is actuated by automatic electrical power, natural forces or human intervention. For example, Theo Jansen's biomorphic kinetic sculptures known as *Strandbeest* give the impression of life as they move in the landscape, because of their apparent self-propelled locomotion. The main feature of their design is a linkage system, a special leg mechanism developed by the artist himself to simulate a smooth walking motion, which is actuated either by wind (in the case of the *Animaris Percipiere*) or human power (the artist himself in the case of the *Animaris Rhinoceros Transport*) (Figures 3.2 and 3.3). When we look at the projects in motion (in the related videos), we are not aware of the chain of mechanical events by which the energy source drives the structures; this may give rise to a sense of animacy, which echoes the artist's claim that he has been "occupied creating new forms of life."[26]

Similarly, Hyperbody's *Muscle Tower II* project presents interesting features that can also contribute to the discussion about animacy.[27] Built in the context of the Hyperbody Research Group by BSc students, the project is a modular and kinetic structure (a prototype for an advertisement tower) that can be assessed in the light of the previous findings. It is made up of a lattice of aluminum rods, interconnected flexibly to each other with hollow galvanized steel spheres. These joints are connected by diagonally placed pneumatic (fluidic) Festo muscle

FIGURE 3.3 *Animaris Rhinoceros*, Theo Jansen, Strandbeest. © Theo Jansen.

actuators to take the shear forces, able to stretch or contract (by air deflation and inflation respectively), thus making the whole structure bend, twist and torque.[28] The inflation and deflation of the actuators that determine the overall shape and motion of the structure are controlled by an air-valve control unit. This comprises 72 control valves, regulated by the Virtools software, which processes sensor input through a MIDI digital-to-analogue and vice versa converter.

The structure can execute one of three behavioral modes, according to sensory input from four motion detectors installed on its feet. When the sensors detect close human presence, the structure responds by bending towards the detected activity for 30 seconds, performing its most active mode – the "response mode." When no external activity is detected, it enters the "sleep mode," performing subtle minimal motions; when distant activity is detected it enters the "alert mode," torqueing and bending slightly around its vertical axis, to suggest that it is looking for someone as if trying to communicate.[29]

It is rather unclear whether, as stated by the project team, the structure can act proactively or it only reacts to external stimuli. Oosterhuis explains that a sudden shift in the motion behavior of structures like this may alternate with less intruding behaviors:

> We do not design buildings like the Muscle and Trans-ports to disturb people, but rather to offer a natural feeling of slowly changing conditions. The building transforms slowly like the weather does. Eventually it may explode like a thunderstorm, but that is only functional as to appreciate the silence after the storm even more.[30]

As the structure swivels and bends, it makes unexpected changes in orientation and motion direction. Although it is programmed to perform on–off reactions to human presence, its overall behavior has been set to demonstrate seemingly unexpected movements, and thus enhance the perception of animacy.

Some recent approaches, however, have attempted to overcome the problem of anthropomorphism in kinetic and computationally augmented spaces. In particular, Konstantinos Oungrinis's and Marianthi Liapi's *sensponsive* logic for computationally augmented and kinetic spaces, although emphasizing the empathic relation between space and users, makes a different case. Following the lines of research factors in autonomous robotics, they explain that "sensponsive architecture targets the emergence of a spatial, assistive 'consciousness' that is seamlessly integrated into human daily environments."[31] A sensponsive system, in a given architectural space, can understand situations by monitoring and perceiving their context, and thus identify the type and intensity of human activities, finally providing a response in a timely fashion according to its intentional modes. The latter are patterns of specific responses that are deemed appropriate for the conditions and the respective context. The system gradually builds a behavior, determined on the basis of optimum responses and previous experience, acquired through feedback. These are manifested by means of flexible spatial configurations, smart materials, speakers and image projections, for spatial and ambient changes in the environment.

The particular sensponsive research project called "Spirit/Ghost," developed in the context of the TIE Lab at the School of Architectural Engineering at the Technical University of Crete,[32] comprises a series of experimental prototypes of mainly educational spaces (such as university lecture rooms), operating in two phases. In the first, the "Spirit" phase, the system comprehends the existing situation by monitoring activities in space, determining context and spatial configuration. The second, the "Ghost" phase, decides on a possible action, thus pointing to a sort of animistic dimension of architecture, while addressing the spatial factors that impact on the students' and teachers' efficiency and performance.[33] But the difference between the senspative approach and other animistic projects, such as the Hylozoic Ground that we will see later, is that it focuses on interior architectural spaces that are transformed in a timely and non-invasive fashion, engaging smooth and discreet responses, to adjust to a given activity, usually educational. Although the system involves operational modes – comprehension, evaluation and intention – that could be regarded as cognitive, it is not intended to be anthropomorphic in behavior, but rather to enhance human activity, and improve the psychological state and behavior of users.

On the other hand, as we saw, ADA and Muscle Tower II embody features that may enhance anthropomorphic ascription and animacy perception – through specific patterns of movement and formal characteristics. Although we can easily assume that structures able to move, react, interact or self-act, may sometimes be perceived as animate, the tendency to see digitally driven structures as "alive" cannot be explained merely in perpetual-psychological terms, because the idea of architecture as a "living organism" has been part of the language and conceptualization of architecture since the 19th century, and lately a recurring concept in the descriptions of intelligent environments and computationally augmented architecture. Despite the physical properties that enhance animacy perception, what counts as living or non-living has certain cultural and historical facets.

The Cultural Aspect

Biological References in Architecture

Architects and architectural theorists have been using biological metaphors since the beginning of the 19th century, with terms such as "circulation," "skin," "structure" or "function," in order to isolate architectural elements from the complex reality of buildings, and render them objective independent categories that can be studied by scientific methods.[34] This tendency owes a lot to the emergence of an epistemological estrangement and weakening of the connection between architecture, art and science, in the 19th century. Although up until then this connection remained faithful to Vitruvian principles,[35] from the second half of the 18th century, and after the industrial revolution, it was radically broken.[36] But at the same time, architects and theorists did not totally abolish the relations between architecture and science. This is evident in the various scientific analogies that they constructed since then to discuss architectonic properties – the most important being the biological and the mechanical ones, which, in the context of the modern movement, were represented by Frank Lloyd Wright and Le Corbusier respectively.[37] In this way they attempted to clarify the principles of a new architecture, either through visual images or metaphors. Wright, for instance, wrote that "Any house is a far too complicated, clumsy, fussy, mechanical counterfeit of the human body…," associating the electrical wiring of buildings with the human nervous system, plumbing with bowels, and heating systems and fireplaces with arteries and the heart.[38] Similarly, Le Corbusier compared the physiology of breathing with the ventilation of buildings, the nervous system with the electricity supply networks, and the blood circulation system with the circulation of people and cars in the city.[39]

Adrian Forty argued that metaphors were used to imply that architecture is not a science and cannot be scientifically analyzed.[40] But Antoine Picon redefined the problem of analogies, which, in his view, are superficial. Science is, recently, more and more considered to be a cultural production with social impact, while technological artifacts historically bear a certain degree of arbitrariness. From this, Picon proposes the permeability of architecture, science and technology, considered within broader cultural frames of interpretation.[41] For example, during the last decades of the 18th century, French architecture evolved strikingly due to the contribution of architects such as Etienne-Louis Boullee and Claude-Nicolas Ledoux, who worked with simple elementary shapes and volumes, and emphasized dynamism and movement (rather than traditional compositional techniques and classical hierarchies). This reflected a new conception of science and a new vision of nature. For scientists and philosophers of the period, science consisted of elements, namely scientific facts capable of being combined in dynamic ways, through analysis and decomposition, while nature was envisaged as a complex set of intertwined dynamic processes.[42]

Picon examined the relation between architecture and the concept and evolution of technological systems. To be able to grasp technological change, besides definitions of technological systems by historians Bertrand Gille and Thomas Hughes, we have to consider the changes of mental states and frames of reference that influence the way such systems are conceptualized, in a given period. Thus, in the late 18th century, French engineers considered a new kind of analytical rather than geometrical rationality, which was linked to a new way of studying efficiency in natural and social processes, and was founded on new relations between the parts and the whole. In this context, a natural phenomenon or a machine was a dynamic composition of parts, emphasizing operations and functions, rather than a static structure configured according to classical rules of order and proportion.[43]

Following Picon here, the biological metaphors used in the language and projects of the 1960s architectural avant-garde, can be considered to be part of a wider frame of reference that comprised cybernetics, and other cultural and epistemological shifts, as discussed in Chapter 1. This is also the case, as we will see, with contemporary conceptualizations of intelligent and interactive architecture, and relates to a historically grounded exchange between machines and organisms. But while the Archigram group emphasized the iconography of such exchanges, Warren Brodey, in his 1967 article *The Design of Intelligent Environments*, used biological concepts such as complexity, self-organization and evolution as inspirational references, to speculate on a possible responsive and intelligent architecture, able to learn from its users, self-act and anticipate behaviors based on acquired experience.[44] This relationship between architecture and life is even more literal today, as the vision of ambient intelligence has led to a rhetoric that describes intelligent environments as "living," "social," "intelligent" or artificial "beings." For instance the author of an article in *Wired* magazine commented on the ability of buildings to mimic living systems by perceiving and reacting to environmental stimuli: "What if buildings could function like living systems… A building that mimics a living system would be able to sense and respond appropriately to exterior conditions like varying winds, temperature swings or changing sunlight…."[45] Maria Luisa Palumbo further argued for a literal analogy between architecture and the attributes of the living body through technology, beyond Vitruvian formal or anatomical analogies: "While the body, invaded and dilated by technology becomes architecture, architecture in turn looks to the body, not as a model of order and formal measurement, but as a model of sensitivity, flexibility, intelligence and communicative capacity."[46] Palumbo explained that, by means of the capacities of electronic technology, information exchange and interactivity, architecture tends to acquire properties of the living body such as flexibility and adaptability:

> The question of sensitivity now indissolubly links the body, machines and architecture. If the distinguishing factor between living and inorganic forms is essentially the capacity to exchange information with the environment and, consequently, flexibility in terms of the capacity to learn and modify, the key innovation of architecture in the second half of the 20th century, characterised by its growing intimacy with machines, is the aspiration to give buildings the sensitivity and flexibility of living systems.[47]

In the same line of thinking, William Braham and Paul Emmons, following Mumford and Giedion, noted that by perceiving our bodies as interconnected systems, we are led to modify our understanding of architecture, which will discover and express our changing subjectivity.[48] Furthermore, Stephen Jones suggested that intelligent environments can be literally considered to be organisms:

> In developing intelligent environments we lose the distinction between organism and environment. The environment becomes an organism because it does all the things that an organism does except, perhaps, self-replication. The kinds of processes that must be operating in the integration of artificial organisms are analogous to those operating in biological organisms. These include complex self-regulatory processes enabled by substantial feedback circuits... These are the sort of things that a brain or nervous system does in response to its earliest experience.[49]

In the following section we will further examine the challenged boundaries between the natural and the artificial, from the perspective of the history of production of artificial life objects. In this way, we will be able to contextualize architecture within a wider set of practices and discourse that regard computational artifacts as what Sherry Turkle has termed "marginal objects."

Exchanges Between Machine and Organism

Although common sense allows us to distinguish and classify living and non-living entities, several computational objects present problems in this regard. Sherry Turkle examined the reactions of adults, children and scientists to the first appearance of computational artifacts and toys that gradually populated American society in the 1970s. According to Turkle, the apparently reactive and at times complex and unpredictable behavior of electronic artifacts, along with their profoundly opaque inner structure – that disabled our understanding of the way they worked – offered no view of their internal workings, and made them appear as psychological rather than mechanical, lifeless objects.[50] This influenced people's thoughts about life and intelligence.[51] Because these artifacts used logic and miniaturized complex chips to perform computations (rather than mechanical components such as wheels, gears, pulleys, springs and pipes), people compared them to another unique logical entity, the human mind. Such perceptions disturbed the psychological boundary between man and machine, the living and the non-living.[52] According to Turkle's findings, people had difficulty in categorizing these objects as either living or non-living. Thus, the fact that they challenged people's psychological boundaries, led her to the idea of the "marginal" object:

> Marginal objects, objects with no clear place, play important roles. On the lines between categories they draw attention to how we have drawn the lines. Sometimes in doing so they incite us to reaffirm the lines, sometimes to call them into question, stimulating different distinctions… Marginal objects are not neutral presences. They upset us because they have no home and because they often touch on highly charged issues of transition.[53]

In the following we will look further into the history of the construction of such "marginal" objects (in early artificial life), in other words the history of practical contestation and redefinition of the boundary between biology and technology. Throughout history, the machine has been closely associated and interchangeable with organism. This history reflects a recurring and constant persistence to challenge and negotiate their boundaries, either by interpreting the organism through the machine or the opposite. Leonardo da Vinci, for instance, in his attempt to give a single interpretation of the biological and the mechanical universe, juxtaposed in his drawings, human body organs or animals with real or imaginary machines. From then on, clockwork human- or animal-like machines, namely automata, especially those of the 17th and 18th century that seemed to simulate the functions of organisms, served as ideological manifestos to demonstrate the conceptual interchangeability of machine and organism.[54]

The construction of automata was grounded on the Cartesian philosophy of mechanism, which extended mechanistic explanations of natural phenomena to biological organisms and even to mental functions.[55] In the second half of the 18th century, however, which was characterized by an emergent uncertainty regarding the validity of mechanistic philosophy, automata entered the philosophical dispute between the mechanistic and the non-mechanistic interpretations of life. They demonstrated the extent to which bodily or mental functions could or could not be reproduced by mechanism. For instance, Jacques de Vaucanson's *Defecating Duck*, which would deceptively simulate a number of internal vital functions, was rather a philosophical experiment to examine the limits of reproducing life and intelligence by mechanical means. As

104 Psycho-Social Aspects

Jessica Riskin explained, this automaton resulted in "…a continual redrawing of the boundary between human and machine and redefinition of the essence of life and intelligence."[56] By showing the inability of its mechanism to simulate living functions, it expressed "…two contradictory claims at once: that living creatures were essentially machines and that living creatures were the antithesis of machines. Its masterful incoherence allowed the Duck to instigate a discussion that is continuing nearly three centuries later."[57]

This constant negotiation of boundaries through machine construction would find its most profound expression in contemporary AI, and the creation of the so-called "sociable robots," such as Cynthia Breazeal's *Kismet* (Figure 3.4).[58] Unlike mechanical 18th-century automata, such robots are endowed with biological and social attributes, and an implied capacity to develop emotional relations with people. This is possible, not only because of their perceived formal characteristics and programmed behavioral patterns;[59] it is also due to our psychological attitude towards computational artifacts, which underwent radical changes since the 1990s. As Turkle showed, although people initially dismissed the psychological nature of machines, favoring instead the uniqueness of the emotional non-programmable nature of humans, later they did not re-establish these "romantic" reactions.[60]

FIGURE 3.4 *Kismet,* Science & Technology, robot, Cynthia Breazeal. © Courtesy Sam Ogden for MIT Museum.

This constant negotiation of boundaries was clearly articulated in the discourse and practices of cybernetics in the 1940s and 1950s. Cyberneticists moved through the track of Claude Shannon's information theory, to suggest that organisms and machines are systems that adapt and adjust to their environment on the basis of the flow and control of a common unit, namely, information. On the one hand, these scientists linked organisms with machines by conceptualizing information as an abstract, immaterial and stable entity that contradicted the instability, uncertainty and chaos of the post-war world.[61] Therefore, homeostasis, as discussed in Chapter 1, was conceived as a common state towards stability that characterized both living entities and machines.[62] On the other hand, the electromechanical devices that cyberneticists built to demonstrate their ideas in real life, strengthened even more the assumption that humans and machines were self-adjusting and environmentally adaptive systems. As Andrew Pickering argued, these adaptive, homeostatic and "conversational" devices, such as Ross Ashby's *Homeostat*, Claude Shannon's *Theseus*, Walter Grey's *Tortoises*, or Gordon Pask's *Colloquy of Mobiles* (discussed in Chapter 4), manifested a "magic" and almost disturbing sense of life.[63] In this way, these devices challenged age-old Cartesian epistemology and ontological boundaries in culture, prevalent in philosophical thinking throughout modernity.[64]

This attempt to synthesize humans and machines within cybernetics was enhanced by the influence of Humberto Maturana and Fransisco Varela's autopoietic theory. The theory expounded the idea that the preservation of a system's internal organization is the determining feature for all things – organic or machinic – to be regarded as living.[65] At the same time molecular biologists, like Erwin Schrödinger, used information theory terms and concepts (such as "code-script," "feedback" and "genetic information") to describe organisms as information processing devices. This was recently demonstrated in the practices used for "mapping" the DNA structure in an information database – the Human Genome Project. These practices led the way towards the conceptual construction of physical-technical objects of knowledge, thoroughly blurring the distinction, as Donna Haraway argued in 1985, between machine and organism, mind, body and tool.[66]

Later, the scientists of Artificial Intelligence attempted to equate the organic with the machinic as well. They either regarded the human mind as an information processing – computational – device, or the human brain as an emergent system, a model for the neural network of the connectionist approach to AI. Both approaches exchanged ideas with cognitive science and neurobiology, as traditional boundaries and distinctions between the natural and the artificial would dissolve; humans and computers were conceptualized as either rule-based devices, or emergent non-deterministic systems respectively.[67] Furthermore, since the 1980s AI researchers have attempted to redefine the concept of life within the AI subfield called Artificial Life (A-Life). Both physical/biological and digital/artificial entities were considered to be living as long as they were constituted by self-organizing complex processes – namely evolution, natural selection, adaptation, learning, metabolism and reproduction.[68] This conceptualization, which enhanced the potential synthesis between biological and artificial life, was not only due to scientific definitions, but also the way A-Life artifacts were represented and addressed (using biomorphic depictions and biological terminology like "organism," "life," and "intelligence"). Thus, researchers erased distinctions between the phenomenal behavior of organisms, and their underlying informational code, constituting, as Hayles put it, "a multilayered system of metaphoric material relays through which 'life', 'nature' and the 'human' are being redefined."[69] At the same time, however, both scientists and the general public were critical of these equations of life, minds and brains with digital A-Life forms and computers. They argued against the efforts of AI scientists to replace human emotion by computation, and emphasized the importance of sensuality and physical embodiment (as for instance in Breazeal's Kismet "sociable" robot).[70]

What seems to be dominant in this historical account of marginal object production is the assumption that the boundary between human and machine is either unbridgeable, as in the romantic reactions that gave prominence to human emotion, or non-existent, as in artificial life and cybernetics in which ontological differences between the natural and the artificial were overridden. In other words, this boundary, although under controversy and dispute (sometimes blurred, sometimes clear cut), was always present. These critiques, as Warren Sack put it, "assign a timeless, unchanging structure to what is better characterized as an on-going struggle to negotiate the ways in which the 'artificial' flows into the 'natural' and vice versa."[71] Instead, Sack proposed alternative critiques, such as Donna Haraway's utopian cyborg myth, which emphasized the permeability of the boundary between artificial and natural, and not what is "essential" and "eternal" about human nature.[72] These critiques make an effort to articulate a vision for subjectivity, to show that human nature maintains continuous relations with the environment, and is part of an eclectic assemblage of heterogeneous, natural and artificial elements.

This potential fusion of the artificial and the natural drives the work of Philip Beesley, associate professor at the University of Waterloo, School of Architecture and principle of PBAI Toronto, as demonstrated, in particular, in his *Hylozoic Ground* project (Figures 3.5 and 3.6). According to Beesley, this installation enhances the empathetic fusion between humans and their environment, and suggests a break with traditional dualisms, as it hybridizes art, science and technology to produce a "responsive architecture," as Beesley calls it, that is opposed to stability and isolation. The installation is able to interact with and adapt to its human visitors, by means of dozens of distributed sensors that detect their presence.[73]

The Hylozoic Ground,[74] which Beesley produced in collaboration with Rob Gorbet, Rachel Armstrong and research colleagues from Toronto, Waterloo, London and Odense,[75] was Canada's entry to the 12th International Architecture Exhibition, la Biennale di Venezia (2010). It is an immersive

Psycho-Social Aspects **107**

FIGURE 3.5 *Hylozoic Ground* detail, Venice Biennale for Architecture, 2010. ©PBAI.

FIGURE 3.6 *Hylozoic Ground*, Venice Biennale for Architecture, 2010. ©PBAI.

plant-like installation that combines a physical modular structure with a computational environment and chemical substances, conceived as an evolving environment that fuses the boundaries between the natural and the artificial, as its functions are supposedly akin to the functions of living systems. It consists of a thin flexible transparent lattice of tens of thousands of lightweight digitally fabricated components that are clipped together, to create lily-shaped ribbed vaults and basket-like columns that stretch and billow. This lattice is covered by rows of mechanical bits resembling fronds, whiskers and flexible tendons that enable the structures to curl in and around the lattice (Figure 3.7). Fitted with a network of cooperating microprocessors (Arduino microcontrollers) and touch

FIGURE 3.7 *Hylozoic Ground*, Venice Biennale for Architecture, 2010. ©PBAI; photo: Pierre Charron.

and proximity sensors, the installation can be triggered by human visitors to simulate breathing, caressing, clamping, and swelling motions, by means of shape memory alloy actuators and kinetic valves. Alongside its mechanized component system, a wet system is engineered to resemble the functions of a coral reef, and the hybrid metabolic exchanges of organisms, like filtering and digesting substances (Figure 3.8). Embedded incubator flasks contain liquid-supported artificial cells that can be stimulated by ambient light and vibrations from the visitors. Primitive glands contain synthetic digestive liquids and salts to absorb fluids and traces of carbon from the environment, thus performing a sort of hydroscopic function.[76]

A quote from the Hylozoic Ground project website reads: "With the Hylozoic Ground project, Philip Beesley is demonstrating how buildings in the future might move, and even feel and think."[77] But unlike Jansen's biomorphic structures that give a perceptual impression of animacy by means of kinetic mechanisms, powered by wind or human force, Beesley's installation is literally

Psycho-Social Aspects 109

FIGURE 3.8 *Hylozoic Ground* detail, Venice Biennale for Architecture, 2010. ©PBAI; photo: Pierre Charron.

endowed with biological functions, such as digestion, breathing and metabolic activities, and, in that sense, it revisits Vaucanson's duck and other 18th-century automata. It is not only capable of perceiving and acting (like sociable robots), but, by means of chemical interactions with its changing environment, it seems to reproduce the very chemical processes that constitute the functions of organisms.

According to Picon, the ubiquity and categorical ambiguity of technological artifacts, evident in biotechnological applications, the multiplication of hybrid artifacts, and the emergence of complex and dense webs of interconnected, highly distributed technologies, has deconstructed the Aristotelian distinctions between the natural and the artificial, creating difficulties in any attempt to put things, either living or non-living, into such categories.[78] As he explains, "the most striking feature of contemporary technological developments is the multiplication of quasi-objects,"[79] namely objects belonging to complex flat networks, defined as pervasive environments, lacking the spatial definition and autonomy characteristic of traditional machines. He thus suggests a conception of technology as an assemblage of heterogeneous components, human and non-human, in disorderly connections.

This modern technology crisis, which Picon uses to interpret contemporary positions in architecture as a response to this new context, is associated to those constantly changing relations between the human and the machine, nature and technology, nature and culture. In this line of thinking, and following Thomas Kuhn's analysis of scientific revolutions and crises, Lister et al. describe the history of artificial life as discontinuous, disrupted by "normal" and "crisis" technologies, analogous to Kuhn's "normal science" and scientific crises.[80] In the current context, "…digital machines represent for us a crisis technology, in so far as all the old stabilities regarding the relations between nature, culture and technology are once again disrupted."[81] Thus, architectural structures, such as those discussed in the first part of this chapter, can be considered to be the products of a crisis that calls for a redefinition and interrogation of the relations and boundaries between the natural and the artificial, user and architecture.

A well-known reason that drives artists and engineers to produce artificial life objects, is the need to understand what is unique about humans and animals, and what separates them from machines, as Bruce Mazlish[82] and Christopher Langton have explained.[83] It is however senseless to claim that the same reason applies to architecture; although certain architectural structures, such as the Muscle projects, present biological characteristics, like motion and interaction, they are not experimental simulations of biological processes, or scientific experiments, like A-Life objects and installations (Hylozoic Ground). Therefore, there must be another reason driving the design of this kind of architecture, which will become evident by examining its socio-cultural dimensions.

The Nature–Culture Divide in Modernity

Since the 1980s, proponents of the Social Studies of Science and Technology, or Science and Technology Studies (STS), have made efforts to expose the hybrid forms in which things are represented, by challenging traditional dualisms that dissociate nature and culture, the scientific and the social, object and subject. For anthropologist Bruno Latour, modernity is a double process of "purification," the separation between nature/science and society/culture, and "hybridization," their mixing. Purification is what moderns pretend to be doing, Latour claims, because nothing is allowed to take place in between nature and society (object and subject), the boundary that defines all reality, although in practice they produce all kinds of nature–culture hybrids (quasi-objects).[84] The modern human accepts these hybrids but conceives them as mixtures of two pure forms, things and subjects or humans and non-humans, which he separates at the same time, in order to extract from them the subject (or the socio-cultural) part and the object (or the natural) part. This distinction is, for Latour, an imaginary construction, because everything takes place between society and nature, in a "middle kingdom" rejected by modernity – a central point of "departure," not separation.[85] Modernity explained everything, but left outside what was in the middle – the production of hybrid technological objects in a post-industrial era of information and "smart" machines:

> ...when we find ourselves invaded by frozen embryos, expert systems, digital machines, sensor-equipped robots, hybrid corn, data banks, psychotropic drugs, whales outfitted with radar sounding devices, gene synthesizers, audience analyzers, and so on... and when none of these chimera can be properly on the object side or on the subject side, or even in between, something has to be done.[86]

A-Life is one of those intriguing practices where the modern subject–object distinctions are redefined. Lars Risan argued that, although A-Life scientists construct artificial "living" beings, at the same time they try to rid them of any subjectivity, because they are considered to be scientific objects of inquiry. Yet, the difficulty to define these distinctions, Risan thinks, following Latour, are due to everyday language, which makes it difficult to draw subject–object boundary lines:

> In our everyday language we – "moderns" – have always been "non-moderns"; "witch doctors"; we do in practice endow our objects with a lot of subjective properties. Unlike, for example, physics, Artificial Life is a technoscience where it is hard to maintain a clear-cut boundary between everyday language and scientific models.[87]

In his text, *Mixing Humans and Nonhumans Together: The Sociology of a Door Closer* (which he signs using the nickname Jim Johnson), Latour discusses the problem of human–machine separation in the case of an automatic door closer.[88] He analyses how this purely technical object is clearly a moral and social agent, an anthropomorphic entity, because it replaces humans and shapes human actions. Latour rejects the separating lines between humans and technological objects that sociologists place; he sees only actors, who are either human or non-human. Such seemingly animate technological objects, social actors in Latour's view, especially apparent in the work of A-Life and the field of sociable robotics mentioned before, challenge modernity's human–machine distinctions.

Lucy Suchman discusses A-Life within the wider philosophical problem of human–machine distinction and the autonomy of the machine: "Having systematically established the division of humans and machines, technological imaginaries now evidence worry that once separated from us machines are rendered lifeless..."[89] She further explains that the insistence on the human–

machine distinction within the modern tradition, drives the prospect of constructing autonomous anthropomorphic machines, in order to make them human-like and "...to be made like us – in order that we can be reunited with them."[90] However, as Suchman points out, although aiming at the opposite, the actual production of intelligent robotic machines lies in the modern tradition of the post-enlightenment era that regards separation and autonomy, rather than relation, as characteristics of humanity.[91]

Bruce Mazlish locates this distinction and need for unification in a historical framework described by three discontinuities – artificial distinctions – in the western intellectual tradition, which were overcome by three great scientists of the past: the first, which placed humans in a dominant, separate position over the cosmos, was overcome by Copernicus, the second, which separated humans from the rest of the animal kingdom, was overcome by Darwin, and the third placed humans over the subconscious (overcome by Freud).[92] Mazlish explains that, as Copernicus, Darwin and Freud refuted these presumed discontinuities, now it is necessary to subvert the *fourth discontinuity*, namely the fallacy that humans are different from the machines they make. Examining the human–technology relationships through Darwinian theory, Mazlish argues that human nature includes both animal and machinic qualities, because tools and machines are inseparable from human evolution.[93]

This is an anthropological view of technology that regards tools and machines as extensions of human organs, an idea that, as mentioned in Chapter 1, goes back to the work of Ernst Kapp, Alfred Espinas and Andre Leroi-Gourhan. Anthropologists of technology see human nature as an evolving identity, unfolding in terms of culture, our "second nature." This is expressed in the form of *prosthetic* devices, either tools or machines – a theme studied by Freud, who called man a "prosthetic god," and Norbert Wiener, who talked about devices like radars, jet engines, and propellers in terms of prosthetic human or animal organs.[94]

Therefore, we can assume that there are cultural factors driving the conception and design of several of the augmented structures and spaces discussed in this book. This is associated with the philosophical discourse and practices of A-Life, and the production of "marginal" objects. Designers endow these structures with behaviors usually attributed to living organisms, and they talk about them as if they are social entities. As A-Life objects and "living" machines challenge human–machine discontinuities, so we can think of digitally-driven architecture as a living machine, or an artificial organism "trying" to subvert Mazlish's fourth discontinuity. Its animate features – motion, proactivity and responsiveness – turn it into a prosthetic extension of humans and human functions (perception, action, intelligence), which echoes the way Oosterhuis conceptualized his E-motive House: "...a social semi-independent extension of the human bodies of the inhabitants."[95]

Conclusions

In this chapter, we addressed the psychological aspects of intelligent spaces and digitally-driven structures. Thus, we questioned the criteria of their conceptualization and implementation. By placing our discussion within the problematic of the modern discontinuity between human and machine, subject and object, we suggested a conceptual framework to explore the psycho-social aspects of digitally-driven architecture and intelligent spaces. We examined the historical grounds of artificial life objects, and the psychological and cultural factors driving their construction. If, as discussed, one of the tasks of A-Life practices is to subvert the human–machine discontinuity, then designing and constructing architecture must be part of this task to "humanize" architecture (considered as an artificial organism or living machine), and to undermine the nature–artifice boundary.

Of course this interpretation concerns mainly experimental examples, such as the Muscle projects or intelligent environments like ADA. Since their designers imbue them with living characteristics and behaviors like perception, action and intelligence, it is possible to argue that they are led by a wider socio-cultural (and perhaps psychological) drive. Like Turkle's "marginal" objects, they blur the boundary between subject and object, human and technology or human user and architecture, thus responding to Mazlish's fourth discontinuity problem.

Yet, these structures are perceived and conceptualized as autonomous entities, separate from the subject, the user. Suchman argued that robotic applications – despite the intentions of AI scientists for the opposite – remain faithful to the modern traditions of autonomy and separation. AI and A-Life engineers produce autonomous entities, which is evident in the way their robots (such as Kismet) are represented in videos, technical reports and webpages.[96] These representations guide the viewer to certain beliefs; they create an archive of social and intelligent populations that mimic humans, thus re-articulating the traditional assumptions about the nature of humans, namely autonomy and separation from machines.[97]

Digitally-driven structures and intelligent spaces, like those discussed previously, only respond to stimuli or make decisions for proactive operations; there is no place for more relational and interactive engagement with users. But in the 1960s, cyberneticists like Gordon Pask and cybernetics-inspired architects like Cedric Price proposed this idea of interactivity and computationally mediated relations between users and architectural space. More recently, architects and architectural historians like Ranulph Glanville,[98] Dimitris Papalexopoulos[99] and Antonino Saggio[100] have also explored this idea. Therefore, in the remaining chapters of the book, we will turn to the concept of interactivity, to examine the possibility to create environments where mutual relations, engagement and cooperation between users and space can unfold.

Notes

1. Nimish Biloria, "Introduction: Real Time Interactive Environments," in *Hyperbody*, 381.
2. Kas Oosterhuis, "E-motive House," *ONL [Oosterhuis_Lénárd]*, accessed June 24, 2014, http://www.oosterhuis.nl/quickstart/index.php?id=348.
3. Kas Oosterhuis, *Hyperbodies: Towards an E-motive Architecture* (Basel: Birkhäuser, 2003), 54.
4. Kas Oosterhuis, "2006 The Octogon Interview," *ONL [Oosterhuis_Lénárd]*, accessed March 1, 2014, http://www.oosterhuis.nl/quickstart/index.php?id=453.
5. "Hyperbody projects," YouTube video, 8:00, posted by "Tomasz Jaskiewicz," July 8, 2009, http://www.youtube.com/watch?v=e5ycPQ2Iy68.
6. MarkDavid Hosale and Chris Kievid, "InteractiveWall. A Prototype for an E-Motive Architectural Component," in *Hyperbody*, 484.
7. See the Hyposurface website at http://www.hyposurface.org.
8. Rem Koolhaas, "The Smart Landscape," *ArtForum*, April 2015, accessed April 4, 2015, https://artforum.com/inprint/id=50735.
9. Linnda R. Caporael, "Anthropomorphism and Mechano-morphism: Two Faces of the Human Machine," *Computers in Human Behavior* 2, 3 (1986): 215–234; Timothy Eddy, Gordon Gallup Jr., and Daniel Povinelli, "Attribution of Cognitive States to Animals: Anthropomorphism in Comparative Perspective," *Journal of Social Issues* 49, 1 (1993): 87–101; Stuart Watt, "Seeing this as People: Anthropomorphism and Common-Sense Psychology" (PhD diss., The Open University, 1998).
10. Caporael, "Anthropomorphism and Mechano-morphism," 219. Also see Jesus Rivas and Gordon Burghardt, "Crotalomorphism: A Metaphor to Understand Anthropomorphism by Omission," in *The Cognitive Animal: Empirical and Theoretical Perspectives on Animal Cognition*, eds. Marc Bekoff, Colin Allen, and Gordon Burghardt (Cambridge, MA: MIT Press, 2002), 9–17.
11. Watt, *Seeing this as People*, 154.
12. See the website of ADA: The Intelligent Space at http://ada.ini.ethz.ch.
13. Kynan Eng et al., "Ada – Intelligent Space: An Artificial Creature for the Swiss Expo.02," in *Proceedings of the 2003 IEEE/RSJ International Conference in Robotics and Automation*, Taipei 14–19 September 2003 (IEEE conference publications, 2003), 4154–4159.
14. Kynan Eng, "ADA: Buildings as Organisms," in *Gamesetandmatch Conference Proceedings* (Publikatiebureau Bouwkunde, Faculty of Architecture DUT, 2004), http://www.bk.tudelft.nl/fileadmin/Faculteit/BK/Over_de_faculteit/Afdelingen/Hyperbody/Game_Set_and_Match/GameSetandMatch_1/doc/gsm_I.pdf.
15. Kynan Eng, "ADA: Buildings as Organisms."
16. Kynan Eng, Matti Mintz, and Paul Verschure, "Collective Human Behavior in Interactive Spaces," in *Proceedings of the 2005 IEEE/RSJ International Conference in Robotics and Automation*, Barcelona, 18–22 April 2005 (IEEE conference publications), 2057–2062.
17. Geoffrey P. Bingham, Richard C. Schmidt, and Lawrence D. Rosenblum, "Dynamics and the Orientation of Kinematic Forms in Visual Event Recognition," *Journal of Experimental Psychology: Human Perception and Performance* 21, 6 (1995): 1473–1493; Judith Ann Stewart, "Perception of Animacy" (PhD diss., University of Pennsylvania, 1982).
18. Fritz Heider and Mary-Ann Simmel, "An Experimental Study of Apparent Behavior', *American Journal of Psychology*, 57, 2 (1944): 243–259.
19. Brian Scholl and Patrice Tremoulet, "Perceptual Causality and Animacy," *Trends in Cognitive Sciences* 4, 8 (2000): 299–309.
20. See Patrice Tremoulet and Jacob Feldman, "Perception of Animacy from the Motion of a Single Object," *Perception* 29, 8 (2000): 943–951; Winand H. Dittrich and Stephen Lea, "*Visual Perception of Intentional Motion*," *Perception* 23, 3 (1994): 253–268.

21 For instance see Viksit Gaur and Brian Scassellati, "A Learning System for the Perception of Animacy" (paper presented at the 6th International Conference on Development and Learning, Bloomington, Indiana, 2006); Scholl and Tremoulet, "Perceptual Causality and Animacy," 299–309.
22 On anthropomorphism in interface design see: Abbe Don, Susan Brennan, Brenda Laurel, and Ben Shneiderman, "Anthropomorphism: From ELIZA to Terminator 2," in *Proceedings of the SIGCHI Conference on Human Factors in Computing System* (Monterey, CA: ACM Press, 1992), 67–70. On robotics see: Brian R. Duffy, "Anthropomorphism and Robotics" (paper presented at The Society for the Study of Artificial Intelligence and the Simulation of Behaviour – AISB 2002, Imperial College, England, April 3–5, 2002).
23 These are autonomous robots that can develop social interactions and learn, as they communicate and cooperate with people. See Cynthia Breazeal, "Sociable Machines: Expressive Social Exchange between Robots and People" (D.Sc. diss., Massachusetts Institute of Technology, 2000); Brian R. Duffy, "Anthropomorphism and the Social Robot," *Special Issue on Socially Interactive Robots, Robotics and Autonomous Systems* 42, 3–4 (2003): 170–190.
24 In their article "The Science of Art: A Neurological Theory of Aesthetic Experience," Vilayanur Subramanian Ramachandran and William Hirstein present a theory of human artistic experience on the basis of the neural mechanisms that mediate it, proposing a list of universal laws that artists, either consciously or unconsciously, apply in their work to stimulate visual brain areas. Caricature is one of those principles, which Ramachandran and Hirstein call the "peak shift" effect. See Vilayanur Subramanian Ramachandran and William Hirstein, "The Science of Art: A Neurological Theory of Aesthetic Experience," *Journal of Consciousness Studies* 6, 6–7 (1999): 15–51. Similarly professor Semir Zeki, discussing the neurobiology of artistic experience in kinetic art, observed that artists downplay the formal and color aspects of their work, in order to enhance the perceptual impact of movement in kinetic objects, and thus maximize the stimulation of the brain area dedicated to the perception of motion. See Semir Zeki, *Inner Vision: an Exploration of Art and the Brain* (Oxford: Oxford University Press, 2000), 179–205.
25 In order to manage anthropomorphism in robotics, we must coordinate the factors that affect its attribution, and thus minimize the discrepancy between real and expected behavior. See Alison Bruce, Illah Nourbakhsh, and Reid Simmons, "The Role of Expressiveness and Attention in Human-Robot Interaction" (paper presented at the IEEE International Conference on Robotics and Automation – ICRA '02, May, 2002); Brian R. Duffy, Gina Joue, and John Bourke, "Issues in Assessing Performance of Social Robots" (paper presented at the 2nd WSEAS International Conference RODLICS, Greece, 25–28 September, 2002).
26 See the Strandbeest website at http://www.strandbeest.com.
27 See "Muscle Tower II," YouTube video, 4:55, posted by "rudintv," July 8, 2009, https://www.youtube.com/watch?v=zoSnof-B4vo.
28 "Muscle Tower II: An interactive and Kinetic Tower," *TUDelft*, accessed July 31, 2014, http://www.bk.tudelft.nl/index.php?id=16060&L=1.
29 Owen Slootweg, "Muscle Tower II," in *Hyperbody*, 404–408.
30 Kas Oosterhuis, "2006 The Octogon Interview," *ONL [Oosterhuis_Lénárd]*, accessed March 1, 2014, http://www.oosterhuis.nl/quickstart/index.php?id=453.
31 Konstantinos-Alketas Oungrinis and Marianthi Liapi, "Spatial Elements Imbued with Cognition: A possible step toward the 'Architecture Machine'," *International Journal of Architectural Computing* 12, 4 (2014): 423.
32 See the project webpage at the website of the Transformable Intelligent Environments Laboratory at www.tielabtuc.com/#/spirit-ghost.
33 Oungrinis and Liapi, "Spatial Elements Imbued with Cognition," 423–428.
34 Forty, *Words and Buildings*, 87–101. For instance, the term "function" used in the new science of biology at the end of the 18th century, referred to the functions

and hierarchical relations of organs. In 1850, Viollet-le-Duc used this term in architectural discourse, to refer to the load-bearing function of tectonic elements. See ibid., 175–176.
35 As Picon explains, this is demonstrated in the professional status of people like Simon Stevin, Francois Blondel, Claude Perrault, and Christopher Wren, who were both architects and scientists. See Antoine Picon, "Architecture, Science and Technology," in *The Architecture of Science*, eds. Peter Galison and Emily Thompson (Cambridge MA/London: MIT Press, 1999), 311.
36 See Forty, *Words and Buildings*, 94–100. Also see Picon, "Architecture, Science and Technology," 314.
37 See the chapters "The Mechanical Analogy" and "The Biological Analogy," in Peter Collins, *Changing Ideals in Modern Architecture, 1750–1950* (Montreal: McGill-Queen's University Press, 1998 [1965]).
38 Quoted in Luis Fernández-Galiano, "Organisms and Mechanisms, Metaphors of Architecture," in *Rethinking Technology: A Reader in Architectural Theory*, eds., William Braham, and Jonathan Hale (London/New York: Routledge, 2007), 266.
39 Ibid.
40 This view initiates from the fact that the success of language metaphors is due to the application of an image from one schema of ideas to another, previously unrelated to the former, thus making a sort of "category mistake." Therefore, as Forty explains, calling an architectural work "functional," which is a metaphor drawn either from mathematics or biology, is based on the initial assumption that architectural objects are *not* biological organisms *neither* mathematical equations, although they manifest the wish to be considered as such. Through this line of reasoning, Forty goes on to argue that the introduction of such scientific metaphors in architecture (either from natural and physical sciences or mathematics), suggests the unlikeness of architecture and science in general. See Forty, *Words and Buildings*, 100.
41 Picon, "Architecture, Science and Technology," 310.
42 Ibid., 316–324.
43 Antoine Picon, "Towards a History of Technological Thought," in *Technological change. Methods and themes in the history of technology*, ed. Robert Fox (London: Harwood Academic Publishers, 1996), 37–49.
44 Warren M. Brodey, "The Design of Intelligent Environments: Soft Architecture." *Landscape* 17, 1 (1967): 8–12.
45 Lakshmi Sandhana, "Smart Buildings Make Smooth Moves," *Wired*, August 31, 2006, accessed August 23, 2009, http://archive.wired.com/science/discoveries/news/2006/08/71680.
46 Palumbo, *New Wombs*, 5.
47 Ibid., 76.
48 William W. Braham and Paul Emmons, "Upright or Flexible? Exercising Posture in Modern Architecture," in *Body and Building: Essays on the Changing Relation of Body and Architecture*, eds. George Dodds and Robert Tavernor (Cambridge, MA/London: MIT Press, 2002), 292.
49 Jones, "Intelligent Environments: Organisms or Objects?" 30.
50 Sherry Turkle, *Life on the Screen: Identity in the age of the Internet* (London: Weidenfeld & Nicolson, 1995), 79.
51 Sherry Turkle is professor in the Science, Technology and Society program at MIT. In the first and second editions of her book *The Second Self: Computers and the Human Spirit,* as well as in *Life on the Screen,* Turkle studied, through several interviews, the psychology of our developing relation to computers and electronic objects when they first became part of popular awareness. She examined how people of different ages interpreted the behavior of "smart" objects, to show the different terms in which they thought about the relations between these technologies and the mind. Turkle explained that technological developments often override their practical uses, and become objects through which people change the way they think about things –

in the case of the computer, what is alive and what is not. But this is not new, because, according to Turkle, it happened every time a scientific discovery was introduced into culture. For instance, the introduction of the clock changed the meaning and perception of time, just as intercontinental transportation changed the perception of distance and communications. See Sherry Turkle, *The Second Self: Computers and the Human Spirit* (Cambridge, MA/London: MIT Press, 2005), 18.
52 Turkle, *The Second Self*, 81, 287.
53 Ibid., 34–35.
54 Aristotle described the concept of the "automaton" in his *Politics*, making the distinction between living, self-moving biological or technological things on the one hand, and inanimate things on the other. Contemporary dictionaries define automata as machines, robots or devices that can perform automatic serial functions, according to a predetermined set of instructions. Such machines have existed since antiquity, both in the western world and the countries of the East. Hero of Alexandria, for instance, in the 1st century AD, described a theater with moving mechanical figures. But since the 14th century, they have multiplied and became more complex, due to the wide use of clockwork mechanisms. See Pamela McCorduck, *Machines who Think* (San Francisco, CA: W.H. Freeman, 1979). For a discussion on whether automata in antiquity simulated biological functions see Sylvia Berryman, "The Imitation of Life in Ancient Greek Philosophy," in *Genesis Redoux: Essays in the History and Philosophy of Artificial Life*, ed. Jessica Riskin (Chicago/London: The University of Chicago Press, 2007), 35–45.
55 For instance, in his book *Man a Machine* (1748), the materialist philosopher Julien Offray de la Mettrie argued that man is foremost a machine. He also argued that vitality, which is a characteristic of all human beings, is only the result of their physical construction, rather than an immaterial force (as in a vitalist perspective). By contrast to Descartes, who separated the material mechanistic body from the mind, for la Mettrie, mental functions were related to the material that all living beings are made of.
56 Jessica Riskin, "The Defecating Duck, or, the Ambiguous Origins of Artificial Life," *Critical Inquiry* 29, 4 (2003): 633.
57 Ibid., 612.
58 Breazeal, "Sociable Machines."
59 These patterns pertain to a series of pre-programmed needs and motives (for instance, the need for social interaction or the need for rest) that the robot tries to proactively fulfil. These motives are connected to respective emotional states that induce different facial expressions ("sorrow," "happiness," "fear," and so on). See Cynthia Breazeal, "Towards Sociable Robots," *Robotics and Autonomous Systems* 42, 3–4 (2003): 167–175.
60 Turkle, *Life on the Screen*, 84.
61 Katherine Hayles, *How we became Posthuman: Virtual Bodies in Cybernetics, Literature, and Informatics* (Chicago: The University of Chicago Press, 1999), 36. Hayles's critique focuses on the efforts of the scientists of the first cybernetic wave to secure that people are in control, and to maximize their capacities, beyond the noise and chaos of an unpredictable post-war world, maintaining their essence as liberal humanist subjects (rational, self-adjusting, free and autonomous individuals), with clearly demarcated boundaries and agency.
62 Ross W. Ashby, "Homeostasis," in *Cybernetics: Transactions of the Ninth Conference* (New York: Josiah Macy Foundation, 1952), 73.
63 Pickering, *The Cybernetic Brain*, 7.
64 Ibid., 22–23.
65 Maturana and Varela, *Autopoiesis and Cognition*, 82.
66 Donna Haraway, "A Cyborg Manifesto: Science, Technology, and Socialist-Feminism in the Late Twentieth Century," in *Simians, Cyborgs and Women: The Reinvention of Nature* (New York: Routledge, 1991), 149–181.

67 Turkle, *Life on the Screen*, 133–136. See Haugeland, ed., *Mind Design*, 2. Also, on the loose and abstract links between artificial and biological neural networks in connectionism see Boden, "Introduction," in *The Philosophy of Artificial Intelligence*, 18.
68 Christopher G. Langton, "Artificial Life," in *Artificial Life: The Proceedings of an Interdisciplinary Workshop on the Synthesis and Simulation of Living Systems Held September, 1987 in Los Alamos, New Mexico*, vol. VI, ed. Christopher G. Langton, Los Alamos, September 1987, 1–47 (Boston MA: Addison-Wesley Publishing, 1989). Artificial Life was established as a scientific discipline in the first conference in Artificial Life held in Los Alamos in 1987. Researchers agreed on four basic principles that artificial organisms should follow, in order to be regarded as forms of life: a) evolution through natural selection, according to the Darwinian definition of life, b) possession of a genetic program (the instructions for its operation and reproduction), c) a high level of complexity in the interaction between their parts, so that their overall behavior would be emergent and indeterminate, and d) self-organization.
69 Hayles, *How we became Posthuman*, 224. Carl Sims's A-Life simulations, for instance, can be considered illustrative of this view. See his Evolved Virtual Creatures website at: http://www.karlsims.com/evolved-virtual-creatures.html.
70 For instance, scientists like MIT's Joseph Weizenbaum argued that AI suggests a flat mechanistic view of human nature. Any attempt to replace human emotional functions (such as interpersonal respect, understanding and love) by programming, is immoral. Cited in Warren Sack, "Artificial Human Nature," *Design Issues* 13, 2 (1997): 58.
71 Sack, "Artificial Human Nature," 64.
72 See Haraway, "A Cyborg Manifesto." Also see Jacques Lacan's theories of subjectivity and Gilles Deleuze and Felix Guattari's concepts of schizoanalysis and "desiring machines."
73 Philip Beesley, Sachiko Hirosue, and Jim Ruxton, "Toward Responsive Architectures," in *Responsive Architectures: Subtle Technologies*, eds. Philip Beesley, Sachiko Kirosue, Jim Ruxton, Marion Trankle, and Camille Turner (New York: Riverside Architectural Press, 2006), 3.
74 Hylozoism is the ancient belief that all matter has life, which, in Beesley's installation, is demonstrated by the intertwined world it projects. Beesley's references about hylozoism include miscellaneous examples, such as human burials in archaeological sites, Wilhelm Reich's philosophy of Orgonomy, Fra Angelico's paintings depicting hybrid worlds, and Manuel DeLanda's essay on nonorganic life. See Philip Beesley, "Introduction," in *Kinetic Architectures and Geotextile Installations*, ed. Philip Beesley (New York: Riverside Architectural Press, 2010), 19–22.
75 See the Hylozoic Ground project website at http://www.hylozoicground.com/Venice/team/index.html.
76 Philip Beesley, "Introduction: Liminal Responsive Architecture," in *Hylozoic Ground: Liminal Responsive Architecture,* ed. Philip Beesley (New York: Riverside Architectural Press, 2010), 12–33.
77 http://www.hylozoicground.com/Venice/team/index.html.
78 Picon, "Architecture, Science and Technology," 324–326.
79 Ibid., 325.
80 Martin Lister et al., *New Media: A Critical Introduction* (London/New York: Routledge, 2003), 326. Thomas Kuhn in his work *The Structure of Scientific Revolutions* (1962) argued that science is a web of institutions, where scientific change depends on social factors (such as generation changes). His central idea is that the development of most sciences is not gradual and constant, but has a typical pattern of interruptions and leaps, and can be divided into historical stages. To describe these patterns, Kuhn proposed three new and related concepts: "normal sciences," "scientific revolutions" and "paradigms." For Kuhn, science is not a stable and cumulative process of

knowledge acquisition, but rather a process in which violent intellectual revolutions are interposed. These revolutions replace a particular conceptual view of the world with another. This view is what Kuhn called "paradigm," a generalized and essential collection of scientific theses and agreements that are shared by scientists of a particular period. The "paradigm" guides the research attempts of scientific communities, and this is the criterion that clearly determines the scientific definition of a field. Therefore, the theories and answers, as well as what is considered to be a problem in the context of a paradigm, changes. During the period of "normal science," which is followed by a "scientific crisis," a universally accepted "paradigm" is adopted. This defines the research problems for scientists, telling them what to expect and supplying them with methods to resolve them. When, after long periods of "normal science," an old paradigm fails to explain an anomaly in a research procedure, and is thus dismissed, alternative theories are developed to replace these paradigms with new ones. This replacement leads to what Kuhn named "scientific revolution": new research problems, methods and anticipated results are settled. See Thomas S. Kuhn, *The Structure of Scientific Revolutions* (Chicago, IL: University of Chicago Press, 1962). Also see John Preston, *Kuhn's The Structure of Scientific Revolutions: A Reader's Guide* (London/New York: Continuum, 2008), 20–21.
81   Ibid., 352.
82   Bruce Mazlish, "The Man-Machine and Artificial Intelligence," *Stanford Electronic Humanities Review* 4, 2 (1995), http://web.stanford.edu/group/SHR/4-2/text/mazlish.html.
83   Langton, "Artificial Life," in *Artificial Life*.
84   Bruno Latour, *We Have Never Been Modern,* trans. Catherine Porter (Cambridge, MA: Harvard University Press, 1993). The quasi object appears in Michel Serres's book *Le Parasite* (Grasset, 1980), in the chapter "Théorie du Quasi Objet." Also, Antoine Picon, in his book *La Ville Territoire des Cyborgs* (Besançon: Editions De l'Imprimeur, 1998), used the term to characterize contemporary technical objects that cannot be clearly defined, because of the loss of their technical autonomy. Bruno Latour, here, examines its ontological aspect.
85   Ibid., 78–79.
86   Ibid., 49.
87   Lars Christian Risan, "Artificial Life: A Technoscience Leaving Modernity? An Anthropology of Subjects and Objects," *AnthroBase*, accessed December 12, 2013, http://www.anthrobase.com/Txt/R/Risan_L_05.htm.
88   Bruno Latour, "Mixing Humans and Nonhumans Together: The Sociology of a Door-Closer." *Social Problems* 35, 3 (1998): 298–310.
89   Suchman, *Human–machine Reconfigurations*, 213.
90   Ibid., 214.
91   Ibid., 213–214. Criticizing robotic artifacts like Kismet, Suchman argues that these machines seem to be working autonomously and proactively, because of the way they are represented in media, which restates traditional assumptions about human nature as autonomous.
92   Mazlish, here, follows Freud who, in his 8th lecture of the Introductory Lectures to Psychoanalysis, given at the University of Vienna between 1915 and 1917, proposed a place for himself among Copernicus and Darwin. See Bruce Mazlish, *The Fourth Discontinuity: The Co-Evolution of Humans and Machines* (New Haven/London: Yale University Press, 1993), 3.
93   Ibid., 8, 216, 233.
94   Ibid., 198. See also Bernard Stiegler, *La Technique and le temps, vol. 2, La Désorientation* (Paris: Galilée, 1996). More recently, Bernard Stiegler argued that human beings are only conceived in relation to the technologies that externalize and make possible their memory, knowledge and cognition capacities, and that evolution only takes place in the assemblages that humans shape with technological artifacts.
95   Oosterhuis, *Hyperbodies*, 55.

96 In Kismet's webpage (www.ai.mit.edu/projects/sociable/videos.html), there are several videos showing the robot interacting with its creator, Cynthia Breazeal, as if it had continuous repeatable skills and a gradually developing capacity for proactive and autonomous behavior. See Suchman, *Human Machine Reconfigurations*, 213–214.
97 Ibid., 238.
98 See Glanville, "An Intelligent Architecture."
99 Dimitris Papalexopoulos proposed possible relations between people and space, the environment and technology (an idea that is further examined in Chapter 4). He considered the significance of the concept of the "quasi-object" in the context of architecture, in order to define the essence of architecture as a "machine of becoming," and to conceptualize architecture as a plethora of possible instances, "virtual" conditions and changing relations. The project, in this case, is not a form, no more conceived as an object; it is conceived as a narrative of relations, in which users, either individually or as collectivities, participate, through interactions, with the object or with each other. Relations then, do not only arise between users and architecture, but also between users themselves. As Papalexopoulos explained, the project is part of an interactive relation "which, in the end, is a relation between users through the artifacts." See Papalexopoulos, *Ψηφιακός Τοπικισμός*, 102.
100 See also Antonino Saggio, "New Crisis, New Researchers: Interactivity vs Transparency, Subjectivity vs Objectivity, Desires vs Needs," *Prima Facoltà di Architettura – Roma La Sapienza*, accessed May 19, 2003, http://www.arc1.uniroma1.it/saggio/conferenze/lo/4d.htm.

4

CONTEXTUALIZING INTERACTIVE ARCHITECTURE

This chapter outlines a theoretical framework for the contextualization of computationally augmented architecture. We start by exploring *interactivity*, a concept that emerged in the practices and discourse of cybernetics and post-war cultural phenomena, and is now frequently used in the context of our contemporary digital culture. We draw on human–computer interaction (conversational, situated and embodied models of interaction), social and post-cognitivist theories of technology (actor–network theory, distributed cognition and activity theory), and theories of technology that use insights from the phenomenological tradition (the post-phenomenological approach to technology), in order to examine the symmetrical and non-symmetrical associations and synergies between people and computationally augmented environments.

We do not aim to include a complete bibliographic review of the related theories, because this would be a task beyond the scope of this book. Instead, we attempt to contextualize our object of inquiry within the main corpus of the theories presented, in order to determine conceptual guidelines for our purposes.

Interactivity

Interactive Architecture

Numerous architects and researchers have been using the term *interactive architecture* in the past few years, to refer to structures, installations or buildings embedded with computational technology. As Michael Fox and Miles Kemp explain in their book *Interactive Architecture*, "the current terminology abounds with terms such as 'intelligent environments,' 'responsive environments,' 'smart architecture' and 'soft space'."[1] They present a variety of projects under this common theme, although their understanding of interactive architecture derives from an overall characteristic, namely the convergence of embedded computation (ambient intelligence), and kinetics – the mechanical motion of building parts. Similarly, the Hyperbody research group at TUDelft addresses the interdisciplinary nature of interactive architecture, i.e. the fusion between architectural design, ubiquitous computing, control systems (embedded sensing and actuation technologies) and computation.[2] According to the researchers of this group, interactive architecture pertains to spaces that can adapt to their ever-changing context in real time, involving spatial, ambient and informational variation of built form, leading even to proactive environments, characterized by dynamic change and immediacy in response.[3] Furthermore, the term "interactive" may refer to buildings that change physically, in response to an ongoing dialogue with their users or environmental stimuli, rather than through linear processes, like those operating in automated buildings (as in a one-way or reactive communication).[4]

Kas Oosterhuis addresses the dynamic aspects of interactivity in architecture: "Interactive Architecture… is not simply architecture that is responsive or adaptive to changing circumstances…." Interactive architecture is "… first defined as the art of building relationships…" involving bi-directional communication between two active parties, which can be either people and/or built components.[5] Oosterhuis and the other researchers at TUDelft seem to understand interactivity as a dialogical process, involving proactive behavior, autonomy and learning through experience, rather than as a process of simple response to stimuli, or exchange of messages between entities. They have applied these ideas in a series of experimental projects, the *Muscle projects* (designed and built in the context of the Hyperbody research group as we discussed in Chapter 1). It is, however, questionable whether the material and mechanical constraints of these projects, as well as their predetermined modes of behavior (such as the three modes of behavior in the *Muscle Tower II* project), engage interactivity in the sense described by Oosterhuis and the other members of the Hyperbody.

Architect Usman Haque – a researcher of interactive systems in architectural design – is similarly concerned with the problematic conceptual confusion between "responsive" (or "reactive") architecture (involving only linear cause–effect operations responding to external stimuli by pre-programmed behaviors) and "interactive" architecture – a distinction illustrated for instance by the way we interact with a cash machine rather than a bank teller with whom we can engage in actual conversation (Figure 4.1).[6] As Haque argues, if we conceptually equate the terms interactive and reactive (or responsive), we are bound to lose a possible fertile conceptual ground.

In the category of so-called responsive architecture, we can place Daan Roosegaarde's series of installations, *4D-Pixel, Liquid 2.0, Wind 3.0, Flow 5.0* and *Liquid Space 6.1*, which can react to human motion, voice or hand clapping, by means of microphones and motion detectors, using pneumatic or electric motor-driven actuators (Figures 4.2 to 4.6). *4D-Pixel* for instance, demonstrates the

Contextualizing Interactive Architecture 123

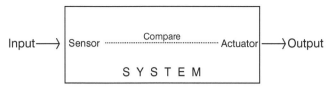

FIGURE 4.1 Diagram of responsive system

aforementioned terminological confusion between interactive and responsive. A kind of "smart skin," as Roosegaarde himself calls it, *4D-Pixel* consists of hundreds of small cylindrical elements – physical "pixels" – which move in and out of a perforated surface, in response to sound frequencies, such as hand clapping. Roosegaarde claims that the project interrelates human activity and architectural form, while adapting to the needs of its users.[7]

A more elaborate approach comes from the work of the Hyperbody. The *InteractiveWall* (2009) was designed and constructed as an installation to showcase a more complex notion of interactivity. It was implemented by the Hyperbody research Group, in collaboration with Festo AG & C (which also commissioned the project), and Burkhardt Leitner constructiv, for the presentation of Festo, at the Hannover Messe 2009 event.[8] According to the designers, "At the heart of the e-motive *InteractiveWall* prototype lies interactivity," that is, an alliance between two participatory active parties, engaging in unpredictable behaviors.[9] The project comprises seven interconnected kinetic components that can bend themselves back and forth, display patterns of light on their skins, and emit localized sounds in response to the presence of people. When active, the installation can react to

FIGURE 4.2 *4D-Pixel,* Studio Roosegaarde, 2004–05. © Daan Roosegaarde.

124 Contextualizing Interactive Architecture

FIGURE 4.3 *4D-Pixel,* Studio Roosegaarde, 2004–05. © Daan Roosegaarde.

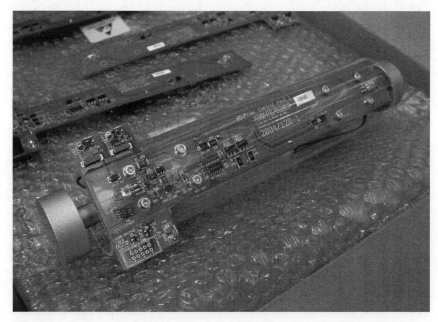

FIGURE 4.4 *4D-Pixel,* Studio Roosegaarde, 2004–05. © Daan Roosegaarde.

Contextualizing Interactive Architecture 125

FIGURE 4.5   *Flow 5.0*, Studio Roosegaarde, 2007–11. © Daan Roosegaarde.

FIGURE 4.6   *Flow 5.0*, Studio Roosegaarde, 2007–11. © Daan Roosegaarde.

the presence of human visitors; each individual component was programmed to deform locally when a visitor approached it (choosing the visitor nearest to it in case it detected the presence of visitors on both sides), taking into account the visitor's distance from the wall, while initiating rapid movement of the light patterns (using LEDs) on its skin. When this happened the other components would attempt to synchronize, by constantly re-adjusting their position, to align with the position of those nearest to them. Sound modalities would also be engaged, ranging from low to high intensity, representing the range from synchronous to asynchronous distribution of the position of the components respectively. The resulting series of complex wave patterns were produced by the interaction between the components' response to the visitor, and the constant synchronization of their position.[10]

Throughout the actual performance, both the installation and the participants would engage in mutual actions, with seemingly unexpected results. Visitors could influence the shape of the piece (albeit constrained by the bounds of its material structure and mechanical configuration), the light patterns and sound emission, and might be physically affected by the result (for instance by stepping back and forth while observing the behavioral changes of the structure). The independent behavior of the three modalities (motion, light and sound), influenced by the incalculable behaviors of the visitors, would result in unanticipated overall patterns – unlike the determinate responses of the *4D-Pixel* that we saw before. The *InteractiveWall* is indeed a technological achievement – a kinetic architectural component with an aesthetically evocative and sometimes surprising behavior. But the interactivity that its designers claim is a perceived one, owing to the complexity of the outcome, rather than a capacity given by dialogical relationships between users and structure. The design team has programmed the piece to be able to respond to external circumstances, and constantly rearrange the relational positions of its individual components (and thus its overall shape) according to a predetermined rule. It is, therefore, on the one hand simply responsive to a single stimulus, and on the other, it is a self-adjusting system, conceptually closer to a first-order cybernetic model. An interactive architecture, however, would mean a communicative and collaborative system, whose outcome is dynamic, unanticipated and co-constructed by both members that participate in the interaction. The system of the *InteractiveWall* does not change its internal state, i.e. its behavioral rules, as a consequence of interaction, nor does it have any proactive behavior, other than a perceived impression of autonomous motion.

Oungrinis's and Liapi's *sensponsive* system, mentioned in Chapter 3, takes the idea of responsive environments one step further. Two general modes comprise the operational logic of the system: the initial *interactive* mode and the more complex *sensponsive* mode. In the interactive mode, users can simply control reconfigurable and kinetic elements, to provide instant spatial changes and modifications appropriate to the given activity. The sensponsive mode, on the other hand, includes cognitive functions such as evaluation, understanding and intention that fine-tune the physical configuration and ambience of space.[11] Although the first mode points to linear action–reaction processes, or what we called *responsive* architecture, the second involves a more complex range of system operations and spatial behaviors.

The project *SpiritGhost 04* in particular, a prototyped proposal for a responsive classroom (following the standard lecture rooms in most university facilities), is engineered to operate on a sensponsive mode, on the basis of five scenarios, manifested in terms of spatial and ambient changes; lecture scenario, board scenario, presentation scenario, discussion and debate scenario.[12] The digital model of the project includes a floor platform, a ceiling infrastructure, and a flexible surface for the ceiling and sidewalls (Figures 4.7 and 4.8). The floor platform consists of parallel strips (further subdivided into seating tiles) that can be elevated automatically, via a scissor lifting mechanism, to vary the inclination of the seating area of the audience. Elastic strips constitute the ceiling and side walls, animated by articulated arms, including magnetic cylinders, spheres and servomotors, hidden in the ceiling infrastructure. These control the curvature of the strips, and the overall flexible fabric attached onto them.[13] What distinguishes this work from other responsive environments is the timely fashion in which responses take place in the sensponsive mode. While projects like *4D-Pixel* only react to specific human stimuli in a linear and predetermined fashion, the *SpiritGhost 04* assumes a much longer relationship with users, one that builds on non-invasive, subtle reactions and optimal responses, on the basis of a set of cognitive features, including understanding of the context, spatial awareness and evaluation of observed activity.

As we shall see, *interactivity* in augmented environments points to a more complex progression of exchanges among entities that are capable of adaptation through learning and memory.[14] The cybernetic concept of *conversation* provides

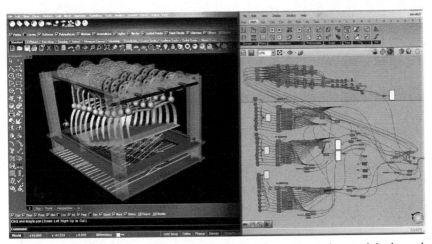

FIGURE 4.7 Parametric control of the *SpiritGhost 04* study model through Grasshopper. © Konstantinos-Alketas Oungrinis.

128 Contextualizing Interactive Architecture

FIGURE 4.8 Project *SpiritGhost 04* visualization of the sensponsive classroom's interior space during different lecture modes. © Konstantinos-Alketas Oungrinis.

a model to understand interactivity as something more than simple response to stimuli or self-regulation. This conversational approach was evidenced in the mutual exchanges that potentially arise in the author's conceptual project *Interactive Wall* (2003) (Figures 4.9 to 4.11).[15] Following this line of thinking, Haque proposes the design of interactive architecture on the basis of open-ended conversations, drawing in particular on the work of cyberneticist Gordon Pask and his "conversation theory" (see also Chapter 1).[16] We will come to this

FIGURE 4.9 Still from a QuickTime movie depicting the *Interactive Wall* in action, *Interactive Wall*, 2003

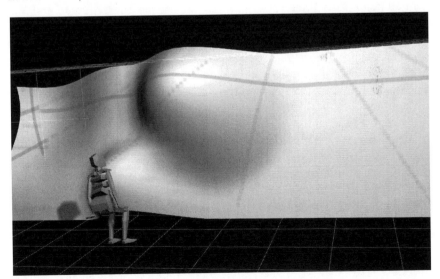

FIGURE 4.10 Still from a QuickTime movie depicting the *Interactive Wall* in action, *Interactive Wall*, 2003

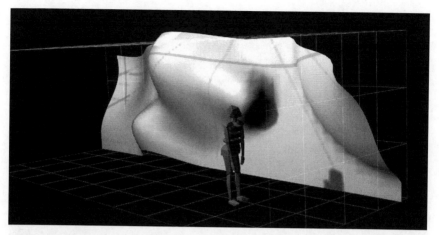

FIGURE 4.11 Still from a QuickTime movie depicting the *Interactive Wall* in action, *Interactive Wall*, 2003

theory and Pask's projects later. But, before that, we will open up our discussion on interactivity, by looking into information technology and human–computer interaction, to trace its beginnings and its current meaning in the context of digital media.

Interaction in the Context of Digital Media

The term *interactivity* generally refers to a mutual activity within human communication or physical phenomena. But in the context of computing, the term pertains to processes of information transaction between people and computers. Thus, it migrates in a space between the natural and the social sciences – which reflects the wider disintegration of distinctions between the physical and the social in postmodern science. But, as we shall see, it is the social connotations of the concept within the context of information technology and human–computer interaction that make it relevant to our discussion, and the conversational and collaborative processes that it implies.

The roots of interactivity as a property of computational systems can be traced in the shift from simple calculating electromechanical machines, using batch processes and punchcard technology, to the first electronic computational systems, developed in the early Cold War period for military purposes in the US (such as the SAGE air defense system), which had the capacity for interactive manipulation of information, through real-time computing and response. These machines led the way for the development of systems with magnetic memory, video displays, computer language, simulation and graphic display techniques. For instance, an early interactive graphical user interface that facilitated one-to-one engagement of users with machines was Sketchpad (1962), which had an immense influence on the development of virtual reality, computer graphics and human–machine interaction.[17] Interactivity is also rooted in the work and ideas of early computer scientists and pioneers, such as Vannevar Bush, Douglas Engelbart and Alan Kay, who explored ideas and techniques for individual access to and management of information in the early post-war period.[18] Thus, the initial technical meaning of interaction meant user control of information; users could modify and override the running operations, in real time, by means of graphical displays, dialogue boxes and menus, so that they could see the results of their intervention in real time.[19]

Interactivity acquired new meanings in the 1990s within the context of computer-based (digital) media. Rooted in Bush's early notion of "associative indexing,"[20] the term meant navigation in websites and databases through hyperlinks, involving user-selected paths, non-linear routes, individual readings and multiple interpretations. Of course, this apparent variability is not endless; it is predetermined by the designer of the interface, the software, the digital medium or the webpage (which run in a limited storage capacity).[21] But, as Pierre Lévy argues, interactivity is not a unique specific property of systems; it is rather linked to methods of communication with a varying degree of interactivity, which is measured by several criteria, such as reciprocity, personalization of messages, telepresence, real-time processing, etc. This degree depends on the capacity of users to re-combine and re-appropriate the transmitted messages (adding nodes or hyperlinks for instance), or to interrupt and reorient the information flow.[22] On such a functional level, *interactivity* refers to the capacity of users to actively interfere with the content of the digital medium, i.e. to place their own information in any form (to change or modify images, texts, sounds, or other forms of information through internet sites such as blogs, forums, bulletin boards, Web 2:0 sites, or open-sim and open-source platforms, such as Second Life and Wikipedia).[23] By means of the appropriate "user-friendly" interfaces, this capacity can be claimed by anyone, not just a few specialists. Thus, when interacting with a computational application, users have the capacity for creative choices and actions beyond the specifications of the designers of the application. Users are no more conceptualized as passive consumers (viewers or listeners) of media and top-down information channels, but rather as active producers programming media content as they wish.[24]

It may be assumed that these capacities are not new, because people always interfered and interacted with printed or analogue media – when for instance they add notes on printed textbooks, or recorded on an analogue audio-cassette according to their preferred sequence of songs. This idea was anticipated by Donald Norman in the 1990s, when he suggested that technologies could be used as external cognitive artifacts for what he called "reflective use." Users could have more control of the artifact because they would be able to compose, manipulate and modify the information provided, as well as reflect on the content.[25] Although we should not underestimate the apparent interactive capacities of analogue media, interaction in digital media is different because the possibilities for information exchange are far more complex, leading to more interpretative responses. Furthermore, user involvement and choice regarding the content of the medium is an intrinsic property of that medium, as well as the on-line structures through which media content is disseminated or sold (through online stores, such as iTunes and file-sharing or "peer to peer" systems such as BitTorrent). This adds another important departure from analogue or older mainstream media.[26]

In that sense interactivity may be analogous to the kind of architectural flexibility which, as mentioned in Chapter 1, provided users with open-ended capacities for creative intervention, according to needs and desires. We need to be cautious though, because interactivity, as described above, does not involve architectural space, but the possibility for information exchange within the functional potential given by digital media. It would be tempting to link the democratic, free and participatory character of the *Fun Palace* with contemporary virtual spaces that also facilitate participation in a public (internet) forum. But since this comparison involves different spaces, one physical and the other digital, we need to articulate it with clearly defined parameters in order to carefully avoid oversimplification.[27]

Of course, as Charlie Gere has suggested, interactivity did not only emerge in the technological developments of the period after the Second World War. It was also a central aspect of wider cultural developments, such as the avant-garde artistic production of John Cage in music, Robert Rauschenberg in painting, the Fluxus movement in performance art, and others. These developments reflected cybernetic thinking and information theory ideas,[28] and explored the role of interaction, audience participation, feedback, unpredictability, interference, indeterminacy and incompleteness,[29] as well as the use of multiple media.[30] Architectural production, such as the work of Cedric Price, for instance, which applied cybernetic ideas in design, was also part of this cultural development. However, in order to explore interaction in the context of architectural space and the nuances of this development, we need to proceed to the examination of the social aspects of interactivity, i.e. human–machine interaction in the complex social framework of daily activity. As we shall see, sociological and ethnographic accounts of interaction in systems of practice within which computational applications are embedded, suggest alternatives to traditional notions of interactivity.[31]

Situated and Conversational Interaction

Interaction, in the context of sociology and communication studies, draws on human face-to-face communication. It is an open-ended dialogue with unpredictable outcome and direction, coordinated through interruptions, corrections and recognitions by participants. In her research on human–machine interaction, Lucy Suchman used observations from conversation analysis, an approach to the study of social interaction pioneered by Harvey Sacks,[32] to argue that human face-to-face communication is not so much a series of alternating actions and reactions (question and response), as a shared and unstructured activity achieved through the continuous participation of its members in conversation.[33]

The mechanisms and structures that make human conversation possible have been studied in the social sciences since the end of the 1970s.[34] More specifically, sociologists and researchers in conversation analysis (a subfield of ethnomethodology) have proposed a series of well-documented rules that organize conversation. This organization maximizes local control in turn-taking in conversation, as well as the direction and development of the subject of discussion, as opposed to other communication mechanisms that impose an a priori and externally dictated agenda (institutionalized forms of dialogue like interviews or lectures). Participants there and then, in the specific situation, decide who talks and on what subject, through the cooperative construction of the course of conversation.[35]

The structure of this kind of communication can maximize the incorporation of unanticipated and unexpected phenomena. Because it takes place in real environments with real people, it is subject to inevitable internal and external interferences (for instance, a misunderstanding, the appearance of a third person or disturbances by natural phenomena), which, however, can be resolved by the participants, who continuously detect and eliminate them. Despite improvisation, the process is coherent and ordered, as conversation and the cooperative construction of turn-taking between participants maximizes sensitivity and adaptation to the indeterminate conditions of interaction. Suchman explains:

> The organization of face-to-face interaction is the paradigm case of a system that has evolved in the service of orderly, concerted action over an indefinite range of essentially unpredictable circumstances. What is notable about that system is the extent to which mastery of its constraints localizes and thereby leaves open questions of control and direction, while providing built-in mechanisms for recovery from trouble and error. The constraints on interaction in this sense are not determinants of, but rather "production resources" for, shared understanding.[36]

This shared understanding and production of meaning in conversation is the basis of interaction. Each action and response not only reflects a participant's intention, but also assumes the other participant's interpretative activity, in order to determine its meaning. Interaction is this contingent process:

> As in human communication, an appropriate response implies an adequate interpretation of the prior action's significance. And as in human communication, the interpretation of any action's significance is only weakly determined by the action as such. Every action assumes not only the intent of the actor, but also the interpretative work of the other in

determining its significance. That work, in turn, is available only through the other's response. The significance of any action and the adequacy of its interpretation are judged indirectly, by responses to actions taken and by an interpretation's usefulness in understanding subsequent actions. It is this highly contingent process that we call interaction.[37]

Using analytical data from face-to-face communication, Suchman addressed the problems that may arise when designing interfaces and human–machine systems, as well as possible solutions. When her book *Plans and Situated Actions: The Problem of Human Machine Communication* was first published in 1987, the common view of most of her colleagues was that electronic objects were interactive almost just like people – albeit with some limitations. This was founded on the wider view shared by most AI and cognitive science researchers, that human intelligence can be modeled by a computer program (see Chapter 2), and that human actions and behaviors are guided by algorithmic rules. But Suchman opposed this idea, arguing for the situated and improvisational nature of human action, i.e. that it is dependent on resources of the immediate and particular situation within which it is performed, and to which it responds.[38] She further objected to the idea that electronic objects are interactive, by running an experiment to analyze and describe procedures throughout the use and operation of a photocopy machine. She observed that the machine could only perceive a small subset of user activities that modified its state (for instance, opening of port, pressing buttons, filling or emptying paper trays etc.). This meant that the potential of the machine to access its users was bounded by certain limitations; at the same time, users had limited access to the operational scenarios and the basic interactive resources of the machine. The result was the detection anew of important asymmetries between people and machines, which indicated the difficulties and the limits in human–machine interaction compared with face-to-face interaction.[39] The continuous contingent resolution of disturbances, and the potential to internalize them in the course of face-to-face communication could hardly be achieved in human–machine interaction: "Because of constraints on the machine's access to the situation of the user's inquiry, breaches in understanding, that for face-to-face interaction would be trivial in terms of detection and repair, become 'fatal' for human–machine communication."[40]

Suchman argued that no interactive machine had been created up until 1987.[41] But this asymmetry still persists; to determine parameters for the design of intelligent systems capable of conversation with humans, professor of design Donald Norman argued recently that humans and machines have no "common ground,"[42] i.e. no common culture or mutually established background for communication. This intensifies their separation and the impossibility for an efficient dialogue: "Successful dialogue requires shared knowledge and experiences. It requires appreciation of the environment and context, of the history leading up to the moment, and of many differing goals and motives of the people involved."⁴³ These requirements significantly obstruct human–machine dialogue, because, unlike people, machines cannot perceive the world and the particular conditions of the environment in which they operate, they have no higher order objectives and, as explained in Chapter 2, they cannot understand the aims and intentions of the people they interact with.⁴⁴ As Norman points out: "But machines and people inhabit two different universes, one of logically prescribed rules that govern their interaction, the other of intricate, context-dependent actions, where the same apparent condition will give rise to different actions because 'circumstances are different.'"⁴⁵

Like Norman, Hugh Dubberly and Paul Pangaro have also argued that what early cyberneticist Gordon Pask called "conversation" demands a common ground, a shared understanding between participants, constructing and reconstructing new meanings throughout, while new messages may be generated.⁴⁶ According to Pask, conversation is the act of concept sharing and conflict resolution during information transfer, a process that can reach coherent agreement between participants. Pask redefined the common sense meaning of conversation, to suggest that a conversational process presupposes organizationally closed and autonomous entities – either human, societal, technological or computational – that retain their identity as they converse. Conversation is "[a] process of conjoint concept execution, by participants A and B, as a result of which the agreed part of the concept is distributed or shared and cannot be functionally assigned 'to A' or 'to B'."⁴⁷ As Dubberly and Pangaro explain, both parties in conversation must commit to engage with each other in dynamic exchanges, break-offs and interruptions, synchronizing and coordinating their actions while negotiating and agreeing on goals for mutual benefit. Participants in the course of conversation evolve and change their internal states (in terms of what they believe, what their views are, and how they relate to others or themselves) as a consequence of learning and memory.⁴⁸

To perform the essential and distinguishing characteristics of conversation, Pask built a number of interactive systems, such as the *Colloquy of Mobiles* installation and the *Musicolor machine*. These projects could respond to what Pask considered to be the most prevalent attribute of pleasurable interactive environments, that is, to "respond to a man, engage him in conversation and adapt its characteristics to the prevailing mode of discourse."⁴⁹ In particular, what Pask considered to be a "socially oriented reactive and adaptive environment,"⁵⁰ the Colloquy of Mobiles, exhibited at the Cybernetic Serendipity show at the ICA in London in 1968, consisted of five rotating suspended figures – the "mobiles" – endowed with the capacity to communicate through audiovisual signals. Three of the "mobiles" were designated as "female" and the other two as "male." The signals were used in the process of satisfying the figures' built-in competitive and incompatible goals, namely hierarchically placed "drives." Represented by two states, an orange and a puce color light (attached on the bodies of the figures), the two "male" drives (which increased their intensity over time), would be satisfied once a light beam of the respective color hit a certain receptor on their bodies. This would be achieved only by "cooperation" with the female figures, which were equipped with reflector items able to send the beam back to the "male" that emitted it. Furthermore, motor-driven rotational and vertical motions of the entities and their reflectors would be engaged to negotiate cooperation and synchronize interaction between male and female, as well as induce "competitive" disturbances against "same sex" figures. The entities could emit specific sounds to signal goal achievement or state of cooperation, as well as learn from their experience and subsequently improve their performance.[51]

This project did not take into account input from human visitors throughout – it was rather concerned mainly with internal interactions. Pask, however, concluded his description of the project wondering whether visitors, by interfering with the installation, would gradually find their own patterns of interaction with it and interpret its signals in a way that they felt would be pleasing.[52] In fact, when the show was on, many visitors interacted with it for a long time while some used mirrors to redirect the beam lights.[53]

The Musicolor machine on the other hand, which Pask constructed along with his colleague Robin McKinnon-Wood in the early 1950s and used in various London theaters, was designed for improvisational performances, involving dynamic interactions between a human musician (usually a piano player) and a light display machine – in what Andrew Pickering called "a dance of agency."[54] Throughout the course of interaction, the system would detect the output properties (sound frequency, transience and rhythm) of the musician's performance, and amplify them as electrical signals before filtering them, using a sound-input analog machine (consisting of uniselectors, capacitors and relays). Once the filtered sound signals exceeded certain threshold values, the machine would produce a variety of colored light patterns. To make the system more complex, these values were circuited to vary in time depending on previous performance and machine behavior. In effect, as Pask noted, the machine was designed to get bored when repetitive actions where performed and shift towards novel activity.[55]

While Pask designed the Colloquy to interact internally within its machine parts (although visitors could engage with the installation with unpredictable outcomes), the Musicolor involved a feedback loop between human and machine, which entailed the musical score, the learning capacities of the machine and its visual display. Here the constant and dynamic interference of the user, i.e. the performer's input (seen as second order cybernetics "disorder"), was an indeterminate factor that played its part in the formation of the overall performance. But not only did the human performance influence the machine output; the latter, with its variable liveliness, was fed back to influence the performer's behavior and so on.[56] Unlike the Colloquy, the goals of the Musicolor were not built into the machine as fixed drives but were rather shared with the human performer, thus thematizing cooperation, suspending any possibility for control over the overall outcome.

Usman Haque addresses the fact that Pask's machines not only reacted to stimuli but they could also change the way they reacted, based on previous behaviors and interactions. They could also take initiative and improvise, in order to encourage further interaction.[57] In order to make this conversational system really interactive, it should be able to cooperate with people through learning, adaptation and negotiation with the produced actions, which is what Pickering calls "a joint product of human–machine assemblage."[58] Along the same conceptual path, architect Ruairi Glynn, who is also a lecturer of Interactive Architecture at the Bartlett UCL, argued that the important aspect of Pask's work is not his technological projects, but rather their conceptual dimensions, which are missing from most contemporary cases of so-called "interactive" art, architecture and design. He thinks that their main disadvantage is the fact that users cannot direct the system beyond the predetermined parameters and behaviors that the designer has "pre-choreographed": "This conversational model is participatory rather than dictatorial and is a form of social communication that promotes a circularity of inter-actions where participants contribute to a shared discourse negotiating their actions and understandings with other participants."[59]

Glynn designs installations and environments that do not simply *react* to human behaviors and environmental changes (as in the case of Roosegaarde's *4D-Pixel*), but instead interact with unpredictable exchanges. His project *Performative Ecologies*, described below, explores the potential implementation of this conversational model.

Performative Ecologies

According to Glynn, *Performative Ecologies* (2008) questions the usual assumptions about intelligent environments, namely that they can only react in a predetermined fashion using pre-choreographed behaviors (Figure 4.12). The project demonstrates the possibility that intelligent systems in space can propose new unexpected behaviors. In this case the architect's role is to design and construct the framework within which the augmented environment will explore and develop its own behaviors and functions adaptively.[60] The installation consists of three suspended robotic entities made of Perspex and aluminum, programmed with a continuous autonomous search mode, namely a basic internal motive to discover ways to attract the attention of the visitors to the installation (Figures 4.13 and 4.14). They are engineered to direct themselves towards the faces of the visitors or towards the other robots. Through continuous interactions (trial and error) and face recognition mechanisms (using cameras embedded on their "heads") to assess the levels of perceptual attention, they are able to determine which of their motions and behaviors are the most successful in terms of attention, discarding unsuccessful actions.

In particular, these robotic entities consist of two basic elements: the "signalers" and the "dancers." The first are light rods, the "tails" of the robots, which initially react mimicking human gestures and movements in order to

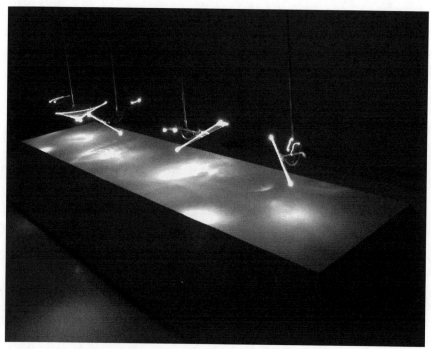

FIGURE 4.12 Ruairi Glynn, *Performative Ecologies*, Emoção Art.ficial, Instituto Itaú Cultural, Sao Paulo, Brazil, 2008. © Ruairi Glynn.

FIGURE 4.13 Ruairi Glynn, *Performative Ecologies*, VIDA 11.0 Exhibition, 2009. © Ruairi Glynn.

FIGURE 4.14 Ruairi Glynn, *Performative Ecologies,* Emoção Art.ficial, Instituto Itaú Cultural, Sao Paulo, Brazil, 2008. © Ruairi Glynn.

learn to execute the most dynamic ones. The second element is the "head" of the robot, which can recognize visitors' faces. By means of a genetic algorithm, these elements work together to change and optimize their motions, in order to successfully draw the visitors' attention.[61] The "dancers" evaluate the fitness of their motions, dismissing unsuccessful ones, and experimenting with more appropriate alternatives. Throughout, the successful motions are stored and combined again to produce new performances. The actions are generated by a "gene pool," a list of possible motions in relation to the algorithm. The mutations that the genetic algorithm is subjected to through this process range in relation to the degree of success of the movements of the robots. If they register a high value of visitor attention, then the levels of mutation rise and the robots become even more experimental.[62]

If no visitors are around, the robots negotiate new behaviors "teaching" each other the successful actions. In particular, by registering the "proposals" of the other two, they compare their "gene pool" (the list of their motions) with these proposals. If they are relatively similar, then they are accepted, replacing a "chromosome" in their pool. If they are very different, then they are dismissed. Communication between the robots is possible by means of a wireless network, although Glynn suggested that they could use their vision system in order to enhance the possibility for interesting mistakes and misunderstandings.[63] As Glynn explains, the installation combines a network of two-way communication and mutual adaptation. Although people and robots are parts of the operational environment and act independently, they negotiate their motions in relation to the other part.[64]

The installation may create the impression of a social animate entity that can interact with people, due to the robots' rapid motions and unexpected orientation shifts, which, as explained in Chapter 3, may induce the perception of animacy. Moreover, these entities can be considered to be examples of *sociable robots*, which, as discussed in Chapter 3, are designed to enable interactions with their users, by incorporating perceptual capacities and a set of pre-programmed motives. Therefore, this project can also be regarded as another case of a *marginal object*, since it can give the impression of a seemingly living artificial entity. Yet, contrary to Suchman's critic of situated robotics, this project is not a reproduction of an autonomous artificial entity modeled on the humanist subject. Glynn implemented an interactive experience that is produced cooperatively by people and technology. Unlike a simple determinist reactive system, the robots' behavior is emergent and unpredictable. Robots participate in experimental performances engaging different combinations of their kinetic parts, and "rewarding" motions that attract the visitors' attention.[65] The fact that the installation has its own specific goals does not contradict interaction as defined previously; this is because interaction entails that participants have and retain their own goals and identity (such as pre-programmed motives), despite the fact that the outcome of interaction may be indeterminate.

Glynn has considered the practical applications of his project; he mentions that it can be a model for buildings that will be able to learn by experience and provide useful domestic services through interaction (such as environmental control and security services). But the most important contribution of the project is that it demonstrates the possibility to create animate environments that operate according to the experience they gain from their interaction and "conversation" with humans. Ranulph Glanville, an architect similarly interested in the potential of cybernetic thinking for architectural design, has also emphasized the importance of interaction for the recognition of intelligence in systems. According to Glanville, the conceptualization of intelligence as a result of the interaction between an observer and the system observed, extends the classical definition of intelligence in the Turing Test. Intelligence is shared and located in the interface between user and machine:

> In effect, intelligence exists in a system in which each component is observer, to itself, and observed, to the other: in which each recognizes intelligence in the behaviour of the other in an interaction, and attributes the quality to the other... Intelligence is... in the actions/reactions shared between the participants, and takes form as their interactive behaviour. Intelligence is shared... We find that intelligence is not what we thought, and that to look for intelligence in some object is pointless. We must look for intelligence between us, as observers, and what we observe, in the space in which we behave together, the space between, the interface.[66]

Having said that, it is now time to review ADA, an installation designed and constructed to induce the feeling of an intelligent entity that interacts with its human visitors.

ADA – The Intelligent Space

ADA, presented at the Swiss EXPO.02 in 2002, was a project developed to explore the potential interaction between human users and computationally enhanced space.[67] The design and implementation of the project involved a multi-disciplinary team of 25 scientists and technicians (biologists, neuroscientists, computer engineers etc.), and another 100 men that worked during construction. It operated daily for 12 hours for more than 5 months, from the 15th of May until the 20th of October 2002 and had 550,000 visitors.

Unlike conventional intelligent environments, which cater for human needs and desires through automation and enhancement of domestic functions, ADA was a temporary entertaining environment, operating for a specific period and number of visitors. Its main goal was to explore the potential interaction between an artificial organism – an artificial intelligence system embedded in architectural space – and human visitors. To achieve this, the designers of ADA examined techniques to make the system able to create the sense of "conversation" with its visitors; it was programmed to detect those that were the most active and responsive to its audio-visual events (that were meant to express emotional states), by constantly evaluating and communicating to the visitors the outcome of its activities. At the same time the system would try to adjust the distribution and flow of visitors. The designers of ADA, referring to its interactive and animate properties, claimed that:

> The degree of success with which visitors can be convinced that Ada is an artificial organism depends on the nature of their interactions. The operation of the space needs to be coherent, real-time, and reliable enough to work for extended periods of time. As well as this, it must be understandable to visitors and sufficiently rich in the depth of interactions so that visitors feel the presence of a basic unitary intelligence.[68]

For its interactive performance, ADA incorporated different sensors and effectors: "vision," "hearing" and "touch." It could trace visitors using 367 pressure-sensitive floor tiles (66cm width with embedded pressure/weight sensor pads and microprocessor), while two sets of three microphones on the ceiling could detect and localize handclaps and simple sounds (through "triangulation") such as the spoken word "ADA." At the same time, it could capture video in real time, using 10 pan-tilt cameras (the "gazers"), which could focus on selective interactions with specific visitors. The effectors consisted of 20 pan-tilt projectors ("light fingers"), which could illuminate selected visitors, colored RGB neon tubes in the floor tiles, a large circular projection screen, and

12 LCD projectors. There was also a ring with ambient lights for atmospheric illumination, while three-color lights in the floor tiles could create local effects.

The control system combined simulated neural networks, agent-based systems, AI of procedural code, and object-oriented software on a computer cluster. Attention was paid to their combination (hybrid model) for their optimal cooperation. In general, to achieve interactive engagement with visitors, the system followed successive behavioral steps: using the "gazers" it detected individuals or teams of visitors and, through the pressure sensitive floor, it communicated this detection by localizing the most active (those who reacted immediately to visual stimuli of the environment). Using information from the pressure sensors on the floor tiles, it determined the position, velocity, direction and weight of visitors. The sound system detected and recognized basic sounds (such as the word "ADA," the pitch of this sound, and handclaps). Depending on visitors' choice, it guided the light heads "pointing" at them and communicating this recognition using local sound effects or local visual floor effects. Thus, by means of a variety of audio-visual stimuli, it encouraged these visitors to group at some spot in the room. When the conditions were appropriate, ADA rewarded them by playing one of its several games.

In particular, to achieve the sense of "conversation" with its visitors, it incorporated six behavioral states (modes), which developed progressively. This procedure involved a predefined number of visitors (25–30) for each session (so that the quality of the experience was ensured and the space could function efficiently) lasting for 5–6 minutes. These behavioral modes were taking place in the following order:

1. "Sleep": Having passed through a series of preparatory spaces, visitors entered the main space.[69] The experience involved smooth blue neon lights, effects of guided lights, simple reactions of the floor tiles and gentle music. As visitors ran and made sounds, the second behavioral mode – "wake" – would be engaged.
2. "Wake": Floor and screen changed their color to bright yellow, while volume and pitch of sounds and music increased. After 24 seconds the system started detecting individual visitors; it would assign a differently colored floor tile to each one, and detect handclaps before creating a halo around the person closer to the source of the handclap. After this transitional phase, the third state was activated.
3. "Explore": During this phase, which lasted for 103 seconds, ADA would trace and evaluate the "responsiveness" of visitors to visual stimuli (to tiles that would blink near them). Highly responsive visitors (those that stepped on the light tiles) were "rewarded" with visual events: the system "circled" them by differently colored blinking tiles, and "pointed" at them using "light fingers," while the projection screen played live video of the event. This succession of interactions would signal the beginning of the

system's source distribution; according to a priority set, the system would pay attention to the most "interesting" visitors, namely the most active and responsive to the audio-visual events.

After a while, the space engaged its "group" mode (which would last almost 33 seconds), with similar interactions but darker colors, and then it would come into the "explore" mode again, but with an essential added behavior: all detected visitors would be prompted by several stimuli to proceed to a specific position in the room. The "game" mode would follow, prompting visitors to "chase" and "capture" an animated floor tile and step on it. After this, ADA would come into the "end" mode with slow music, darker lighting and traveling red waves on the floor, prompting visitors to exit the room.[70]

ADA was programmed to follow a set of prioritized motives, aiming at the optimal outcome. The aim was to maximize the parameter H representing the potential "happiness" of the environment. The value of H was a function of a number of programmed behavioral motives, "survival," "detection," and "interaction." "Survival" was a measure of how well ADA fulfilled its basic requirements, that is, the preservation of a certain flow of visitors in time and a determinable average degree of velocity; "detection" would determine how well the environment could trace and collect data from people (who would be a precondition for the progress of interactions); and "interaction" would count the number of successful interactions with the visitors, assigning higher values to more complex interactions, such as team games.[71]

ADA was radically different from conventional intelligent environments (see Chapter 2), which are mainly domestic environments catering for user desires and lifestyles by means of ubiquitous adaptive and proactive computing. ADA was a temporary installation constructed to demonstrate the potential of advanced artificial intelligence to interact with people in space through play, as well as explore people's perception of an artificial non-anthropomorphic entity. Indeed, the project designers claim that this interactive and collaborative experience influenced visitors' perception of ADA as a living intelligent organism:

> [...] the users of the building must engage in a dialogue with the building in order to achieve their goals, rather than simply controlling a few parameters. In the course of this dialogue, the users begin to treat the building as a single entity, and attribute purposeful cognition to the processing underlying the building's actions. From the viewpoint of the user, the building becomes more than the sum of its shell and subsystems; it appears to be alive.[72]

Certainly, as explained in Chapter 3, this attribution may be due to the perceived intelligence induced by ADA's *opaque* behavioral mechanism. Yet, it was the interactive engagement that made this project interesting. As described above, the visitors would gradually participate in a relationship with the environment, having the impression that they converse with it and that they collaborate in the production of their experiences. They would gradually learn to play with the environment, while the system would locate and recognize the visitors' behavior, responding accordingly for the development of interaction. However, this improvisational experience of "conversation" cannot be compared with the conversation between humans, because face-to-face communication involves intentional subjects. Although ADA had a specific aim and intention (its degree of "happiness" represented by the value H), it was pre-programmed as an internal motive that determined the range of its actions and interactions. Therefore, interaction was asymmetrical; ADA's agency was a delegated one (an idea that we will examine later), namely the implementation of its designers' intentions. Furthermore, although visitors could freely move in space, interaction was limited by the programmed motive and the operational rules of the system. If the visitors had been able to modify these rules by accessing the system's functionality, a more unexpected series of behaviors and interactions might have occurred. This prospect, as we will see in the next chapter, is the project of "tangible computing" and end-user driven intelligent environments.

In both ADA and Performative Ecologies, the prospect of creating flexible adaptive architecture, which, as explained in Chapter 1, relates to the potential for creative user participation and intervention, remains a question. In such an environment users can cooperatively propose and create often unanticipated behaviors and functions. Therefore, in the following we will explore this potential, taking into account the asymmetry of human–machine interaction discussed by Suchman and Norman. We will draw on theories that study the relations between humans – the social – and the material-technical world; we will look into interdisciplinary perspectives of human–machine interaction, and theories that address non-human agency, as well as symmetrical and asymmetrical socio-technical assemblages.

Symmetry and Asymmetry in Human–Machine Interaction

In the rest of the chapter, our examination of human–machine interaction will be informed by theories of cyberculture, actor–network theory and post-cognitivism. We will address concepts such as distributed cognition, shared and emergent intelligence, and interchangeable agency. We will discuss social and post-phenomenological theories of technology, activity theory and theories driven by insights from phenomenology, such as the theory of embodied interaction.

Human–Machine Agency: From the Cyborg to the Posthuman

As an iconic image of the interchangeable relations between humans and machines, the cyborg (short for cybernetic organism) represents the multiple connections between nature and artifice, radically challenging humanist agency.

Although the idea has always been a mythical and uncanny figure in fiction,[73] the term is rather new; Manfred Clynes and Nathan Kline, who first coined it in their text "Cyborgs and Space" (published in *Astronautics* in 1960), described it as a technologically augmented human organism capable of surviving extraterrestrial environments.[74] In everyday language the term describes different cases of human–machine interconnection in industrial and post-industrial societies: in military industry, medical research and biotechnology, computer science, toy and entertainment industry and human–machine interface design.[75] But after the publication of Donna Haraway's seminal text "A Cyborg Manifesto,"[76] the cyborg acquired connotations beyond science and technology, and resonated with political theory, military history, critical theory, computer science, medical sociology, psychology and cultural theory. Zoë Sofoulis, examining the impact of Haraway's text, observed that it has influenced a great deal of technological and cultural phenomena:

> [...]from mundane computer and videogame interactions, or human interactions mediated by computers and the Internet, to more extreme examples of anything involving physical and virtual intimacies between humans and machines, especially where the latter were seen to exert some type of agency. This included artworks, involving bionic bodies such as Stelarc's third arm and more recent gestures towards the "posthuman," or interactions between humans and nonhuman elements (such as interactive artworks in digital media or installations fitted with responsive sensors).[77]

To explore alternating assumptions about subject–object and human–machine relations, in the following we will look at three key issues associated to the cyborg concept: the posthuman; agency as a property of the machine, i.e. the technological part in the construction of the cyborg; and interaction between human and non-human things as discussed in actor–network theory.

From Autonomous to Relational Systems

Apart from a metaphorical myth about human–machine relations, Haraway's cyborg is a proposal for a new subjectivity, a fictional construction of fragmentary assemblages and hybrid interconnections between humans and non-humans that maps our human, social and bodily reality: "By the late twentieth century, our time, a mythic time, we are all chimeras, theorized and fabricated hybrids of machine and organism; in short, we are cyborgs. The cyborg is our ontology; it gives us our politics."[78]

For Haraway, the cyborg, in the era of postmodern technoscience and biotechnology, is the personification of a future that challenges and deconstructs traditional dualisms of Western culture (such as mind–body, technology–nature, subject–object, and culture–nature), where, according to the established feminist critic, the first dominant-masculine part suppresses the latter.[79] Haraway (along with Bruno Latour and other proponents of actor–network theory as we will see later) argued that these dualisms fade out as we conceive of ourselves as parts of a world of socio-technical entities.[80] Therefore, the cyborg is not necessarily a literal human–machine system, but is located on the boundary between the human and this historically constructed subjectivity – the posthuman. In her book *How we Became Posthuman*, Katherine Hayles explains:

> [...] it is important to recognize that the construction of the posthuman does not require the subject to be a literal cyborg. Whether or not interventions have been made on the body, new models of subjectivity emerging from such fields as cognitive science and artificial life imply that even a biologically unaltered Homo sapiens counts as posthuman.[81]

In the context of first order cybernetics, information was conceived as a dimensionless and immaterial entity, independent of its physical substrate, which has been historically connected to the decomposition of the bounds of the liberal humanist subject, leading to the posthuman.[82] For posthumanism, the natural substrate, the body, disintegrates and is replaced by an amalgam, a collection of heterogeneous parts, a material and informational entity, whose limits are subject to continuous construction and reconstruction:

> Because information had lost its body, this construction implied that embodiment is not essential to human being. Embodiment has been systematically downplayed or erased in the cybernetic construction of the posthuman in ways that have not occurred in other critiques of the liberal humanist subject, especially in feminist and postcolonial theories.[83]

Disembodiment and the degradation of the body draw on a cultural dualism, particularly dramatized in the central theme of cyberpunk science fiction, namely the dream to escape the mortal materiality of the body by uploading consciousness and the mind in cyberspace.[84] Yet, in other cases, the posthuman is not conceived as a disembodied, transcendental and immaterial cyber-consciousness, but as a larger, extended and technologically augmented system, whose biological core is the physical body. Drawing on a Heideggerean conceptualization of human bodies, Mark Poster argues that our bodies are not just in-the-world, but extended "through the infinite wires and radio waves that criss-cross the planet continuously and in ever-increasing density" and draped over the whole planet through networks.[85] This fusion of human bodies and electronic networks recalls what media theorists and cyberneticists observed some decades ago; Marshall McLuhan, for instance, suggested that his "global village" was an electronic prosthetic extension of the human nervous system,[86] while cyberneticist and anthropologist Gregory Bateson questioned traditional notions of human interiority through the multiple channels of communication with which humans exchange information with the world.[87] Both echo William

Mitchell's argument, as discussed in his book *Me++: The Cyborg Self and the Networked City*, that the integration of the built environment with the various networks of movements and flows that make up the electronic space of cities, extend and artificially enhance the human muscular, skeletal and nervous system.[88] Mitchell's and Poster's definition of the posthuman is different from Hayles's, as it presupposes the centrality of the agency of the biological body for the construction of the posthuman. Mitchell says: "We will do better to take the unit of subjectivity, and of survival, to be the biological individual *plus* its extensions and interconnections."[89] In the following, however, we will attempt to demonstrate the interchangeable agency in human–machine interconnections.

Discussing the question of agency in interaction, Lister et al. point out that "…there is no historical constancy in the human user being the master, and the non-human technology, the servant."[90] They argue that historically only one of two parts, either human or technology, is the prevailing actor in interaction:

> In many ways, arguments about interactivity replay those concerned with whether technology or human, social purposes form the principal historical actors: whichever way we argue these cases, we conclude that one, and only one, component of human–machine relations contributes action to history, while the other plays a merely supporting role, and is acted upon by the active one.[91]

Since agency can be attributed not only to humans but also to technology,[92] interaction could be regarded as an ideal attempt to resolve human–machine separation through a more collaborative approach. Several multimedia installations and performances have demonstrated this collaborative conception of our relations with the machine, and the exchangeability of agency. For instance, since the 1990s, Australian multimedia artist Stelarc has used prosthetic devices and interface technologies to explore alternative human–machine relations, pushing the limits of the body beyond its biological functions. For Stelarc, the body is insufficiently equipped to face the quantity and complexity of received information, or the accuracy, speed and power of technology. Following the lines of cyberpunk discourse, and its emphasis on the degradation of the material body, he argues that the body is "obsolete" and that it can be replaced by technological applications.[93] In his *Fractal Flesh* internet broadcasted performance, Stelarc's body incorporated a prosthetic robotic third hand, while muscle stimulation systems were activated by data coming from collective internet activity.[94] The prosthetic hand was indirectly controlled by his involuntary sensor-recorded muscular activity and body motions (Figure 4.15).

Some critics, though, suggest that Stelarc's performances hardly represent his belief that the body is obsolete, because they demonstrate collaborative and fluid relations with technology. His performances enact the body's open potential for new forms of connectivity and interaction. John Appleby, for instance, argues that,

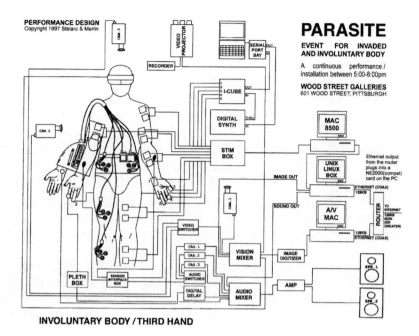

FIGURE 4.15 Stelarc, *Fractal Flesh,* Wood Street Galleries, Pittsburgh, 1997, Diagram – Stelarc. © Stelarc.

Rather than exhibiting the obsolescence of the body, Stelarc's Internet performances demonstrate the importance of conceptualizing technological development in terms of bodily interaction, inasmuch as they open up new spaces for encounters between the human biological form and technology.[95]

Appleby thinks that Stelarc's *Third Hand* performance[96] showcases the fluid nature of the boundaries of the self (Figure 4.16). It poses the question whether the robotic prosthetic third hand is or is not part of the body, and whether the relation between human and machine is instrumental. The fact that the prosthetic hand (which does not replace the right natural hand, but is added to it) is controlled by the abdominal muscles of the artist and not by the natural hands, makes it difficult to perceive it as a tool. It is rather perceived as an extension of the body. At the same time, the fact that his real left hand is activated and controlled externally (and not by the internal nervous system of the artist) showcases the shift of control between the human body and its technological equipment.[97] As Mark Poster has also suggested, Stelarc's performances do not reproduce the humanist subject of autonomous action exercising intentional dominant power on nature and the machine. Unlike this instrumentalist attitude to technology, his performances suggest a fundamental reshaping of the relations between human and nature, body and mind, individual and society, human and machine.[98]

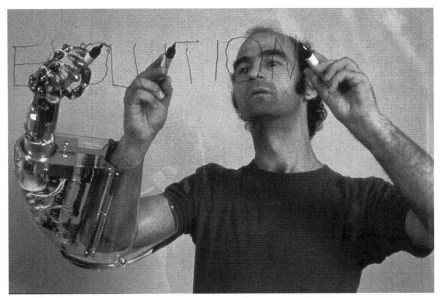

FIGURE 4.16 Stelarc, *Handswriting*, Maki Gallery, Tokyo, 1982. Photographer – Keisuke Oki. © Stelarc.

This understanding of Stelarc's performances suggests a more integrated conceptualization of the cyborg. Although, as philosopher Andy Clark has argued, we have always been cyborgs because we have always been able to incorporate tools and supporting cultural practices in our existence,[99] the embodied technologies that are increasingly well-fitted on our bodies, create a radical possibility for a more intimate, yet natural, human–machine merger. Kevin Warwick's *Project Cyborg* experiments, for instance, have radically implemented this potential by means of surgical implant technology.[100] In this case, the cyborg does not simply represent a prosthetic attachment of technological components onto biological parts, exchanging agency, but a dramatic invasion of technology within the natural body operating beyond direct consciousness. Biotechnological artifacts, transplants, minimized implants and genetic engineering applications are manifestations of our possibility to seamlessly incorporate technological apparatuses in biological substances. They demonstrate the foundational argument of cybernetics, namely the radical coupling between biology and technology. What is emphasized here is the unstable and perpetual boundaries between biology and technology, as well as the fact that the cyborg is a complete entity, equally constructed by inseparable biological and technological components. For Haraway and Latour, the products of technoscience, biomedicine and biotechnology are natural-technological objects or material-semiotic actors, i.e. the result of interactions between their material dimension and their semiotic or technical functions. Thus, as Sofoulis argues following Haraway, technology presents epistemological and political potential for developing alternative interconnections with the "other" and the living world.[101] As she further explains, technology, in Haraway's work, is opposed to the stereotypical feminist view, linked to rationalist, military and ecological origins in capitalist western society: "…the Cyborg Manifesto acknowledged the pleasures and desires we hold in relation to the nonhuman entities that are part of our life-world, and the socio-technical and material-semiotic hybrid entities and plural identities we might form with nonhumans."[102]

Although the figure of the cyborg provides the means to analyze different relations between bodies and objects, it remains an autonomous, discreet and individual entity (albeit hybrid) because, as Hayles suggests, the human is maintained as the center of posthumanism. But, according to Suchman, this obscures the possibility of the cyborg to override its boundaries in everyday instances and transform into a complex field of distributed socio-material interconnections:

> The past twenty years of scholarship in the humanities and social sciences have provided new resources for thinking through the interface of humans and machines. Reappropriation of the cyborg, as a figure whose boundaries encompass intimate joinings of the organic and the inorganic, has provided a means of analyzing myriad reformations of bodies and artifacts, actual and imagined. Expanded out from the singular figure of the human–machine hybrid, the cyborg metaphor dissolves into a field of complex sociomaterial assemblages, currently under study within the social and computing sciences.[103]

Suchman observes that, along with Haraway's feminist model of the cyborg, which has informed us about the political consequences, the shifting boundaries and the transformational potential of human–machine assemblages, it is time to move on with particular configurations and their consequences: "… we need to recover the way in which more familiar bodies and subjectivities are being formed through contemporary interweavings of nature and artifice, for better and worse."[104]

In the last 30 years, a developing body of theories examined the socio-material practices that conceive agency in terms other than its humanist perspective. The effective agency of subjects, as Suchman suggested, does not result from the reproductions of autonomous machines modeled on the human subject (as in the case of robotic A-Life entities), but in networks of social and material relations and configurations, in the complex assemblages of people and things in a particular place and time.[105] Suchman locates such assemblages in several new media art performances that seem to implement the idea that experiences of interaction are reproduced by machines or environments that recognize, feel or react to our behaviors through collaborative processes. She discusses interactive media installations, such as the *TGarden*[106] or Stelarc's *Prosthetic Head*,[107] to demonstrate the dynamic potential of digital media to redefine humans and machines as well as their intermingling. She explains that such forms radically challenge the traditional image of the human, the liberal humanist subject, more than the most ambitious projects on humanoid robots. She adds: "Central to this innovative approach is abandonment of the project of the 'smart' machine endowed with capacities of recognition and autonomous action."[108]

Thus, agency and autonomy should characterize neither humans nor machines, but their interactions. Designers and engineers of humanoid machines in A-Life aspire to mimic the autonomous agency of humanist subjects, and produce robotic entities with human capacities in order to challenge their status as inert objects. Suchman, alternatively, suggests that agency and autonomy should be seen as the consequence of the continuous socio-material practices that govern the relations between humans and machines.[109] As Bruno Latour has also argued, the products of actions in socio-material networks are not to be found in a "pure" form but rather as hybrids of scientific-technological objects and social phenomena and entities.[110]

Agency in the Socio-material Actor–Network

Bruno Latour, John Law and Michel Callon proposed a theory in the context of the Social Studies of Science and Technology, aka Science and Technology Studies (STS), to articulate a plan for the description of the structure of the processes pertaining to the construction of hybrid forms of coexistence and interaction between society, science-technology and nature. Rejecting essentialist notions of culture, nature, society and technology, they introduced the so-called Actor–Network Theory (ANT) to argue for a socio-technology that overrides distinctions between the social and the technological, subjects and objects.[111] This theory attends to socio-technical networks engaging actors, who, instead, may be called actants,[112] to stress their possible non-human status,[113] caught up in a heterogeneous web of symmetrical associations. According to Latour, in this framework, sociology cannot be defined as the "science of the social," but as the tracing of another matter made of social relations.[114]

Although the theoretical consistency and even the very name of ANT are disputed,[115] part of its appeal seems to be its versatility and its capacity to resonate with diverse areas. Considered as a hard-to-grasp mode of thinking, a working method rather than a theory, ANT can trace relations and symmetrical interactions between agents (entities, things and people); these are nodes with no stable properties, defined only in relation to other entities.[116] Michel Callon explains: "…not a network connecting entities which are already there, but a network which configures ontologies. The agents, their dimensions, and what they are and do, all depend on the morphology of the relations in which they are involved."[117] Therefore, in the framework of ANT, social action is not the result of individual or collective purposeful human action, and never takes place in individual bodies. It is foremost relational because it mediates between humans and the increasing quantity of non-human things in the cultural landscape. John

Law explains: "If human beings form a social network it is not because they interact with other human beings. It is because they interact with human beings and endless other materials too... Machines, architectures, clothes, texts – all contribute to the patterning of the social."[118]

Converging with Haraway's cyborg anthropology, ANT overrides the instrumentalist view of technology, within which artifacts are neutral intermediaries of human actions.[119] Instead, actor–network theorists propose the idea of *mediators*, arising in the complex relations between different heterogeneous actants, which are constantly modifying meanings, making possible "programs of action"[120] that point to metaphors, synergies and coalitions – namely "delegations," "inscriptions," "de-inscriptions," "translations" and so on.

In his article "Where are the missing masses?," Latour examines how action within actor–networks is negotiated between humans and artifacts, to argue that artifacts, although missing from studies of sociology (thus being the missing masses), are always present in the socio-technical relations that they mediate: "To balance our accounts of society, we simply have to turn our exclusive attention away from humans and look also at nonhumans."[121] In this article, Latour describes how programs of actions may be *delegated* from one actant to another, either human or non-human. These programs of action take the form of *scripts*, i.e. "instruction manuals" that include those visions embedded in the product for its intended use and meaning, involving both physical properties and socio-technical aspects.[122] According to Latour, scripts are thus delegated from less reliable to more durable actants, from written instructions or humans to artifacts and technologies. In the framework of ANT, this *translation* of action from one actant to another is called *inscription, transcription* or *encoding*.[123] Social action takes place in a network of heterogeneous relations; it is the result of the actions in that network, which is activated by human and non-human things: "The argument is that thinking, acting, writing, loving, earning – all the attributes that we normally ascribe to human beings, are generated in networks that pass through and ramify both within and beyond the body. Hence the term, actor–network – an actor is also, always, a network."[124]

Non-human things in actor–networks could include various technologies; from medical research and intervention techniques, to telecommunication or transportation systems, and genetic technologies, which enable these interactions and interconnections. For actor–network theorists, the mediations, delegations and agencies that develop in the assemblages, networks, and interactions between humans and technologies, are symmetrical. Such is the case, as we will see below, with the post-cognitivist strand in cognitive science, studying cognitive processes in larger cognitive systems that comprise social and physical environments, material artifacts and cultural entities, beyond the bounds of the human brain.

Distributed Cognition

During the last decades of the 20th century some scientists and theorists in cognitive science and the theory of mind, examined the complex inter-dependencies between people, objects and technological systems (which are often ignored in traditional theories), in order to suggest the potential extension of the unit of analysis of cognitive processes from individual solitary agents to broader collective systems.[125] In particular, Edwin Hutchins developed the theory of distributed cognition in the mid-1980s to argue for an embodied and socially distributed cognition, and was thus able to explain cognitive phenomena that emerge in interactions between social entities, mental structures and material artifacts in broader socio-technical real-world settings (like the airline cockpit). Intelligence, in the framework of distributed cognition, is not an abstract ideal property located in a single mind-information processor, but an emergent property of interactive loops between social entities, mental structures, material artifacts and technological objects, involving perception and action in a real environment.[126]

Distributed cognition theorists emphasize symmetry among the nodes of heterogeneous networks. Like actor–network theorists, they suggest that agency is a property of large systems that engage rich interactions between people, physical artifacts and representations. These theorists look further than the idea of a cognitive system extended into the material world, such as that proposed by Andy Clark and David Chalmers in their seminal work *The Extended Mind*.[127] They argue that distributed cognition includes structures of the material, social, cultural and historical environment that participate in cognitive processes.[128] They still use concepts from the traditional theory of mind, but in larger socio-technical networks, organized in a cognitive system through the functional relationships of the elements that participate in it, rather than their spatial co-location. In this way, distributed cognition theory can explain the processes of collective and cooperative problem solving, how knowledge is disclosed, the mechanisms of communication synchronization and the cooperative function of communication.[129] In the context of human–computer interaction, Donald Norman discussed this idea in his book *Things that make us Smart* (1993), where he argued that people in everyday work activities (involving interaction with artifacts and cooperative processes with others) can distribute and extend their memory and reasoning abilities to the external physical and social world. Distributed intelligence is tightly coupled to the world, and takes some of the cognitive burden off humans, by means of artifacts that can expand our memory and computational skills.[130]

As we saw, ANT theorists are concerned with heterogeneous networks characterized by symmetrical associations and interchangeable agencies between actants; other entities might act and others might delegate agency to other entities. This emphasis on associations has subjected ANT to several critiques, mainly because the theory fails to account for what is outside associations, like the social, cultural and historical context that may shape and differentiate between networks, as well as the fact that humans, unlike things, are capable of intentions and can pursue their own interests.[131] Scholars like Andrew Pickering have criticized the concept of symmetry suggested in ANT, arguing that a key distinction between human and non-human entities is intentionality. Although Pickering admits that action and agency are emergent attributes of networks in which humans and non-humans interact, the relations of humans/non-humans, and the very construction of networks relies on the motivational intentions of human actors.[132] This *asymmetrical* view of human–machine relations has been articulated by a number of theories driven by insights from the phenomenological tradition, and committed to the primacy of technology-mediated intentional human subjects.

Activity Theory: Asymmetrical Relations

Two major texts, published at the end of the 1970s and the beginning of the 1980s, introduced Activity Theory to the international audience: the English translation of Aleksei Leontiev's *Activity, Consciousness, and Personality* (1978), and a collection of texts by Leontiev and other activity theory scholars edited by James Wertsch (1981).[133] After 1990, the number of people interested in the theory increased significantly, especially in the field of Human–Computer Interaction. Activity Theory was initially developed around the 1930s by soviet psychologists Aleksei Leontiev and Sergei Rubinstein on the basis of Marxist philosophy. It was an attempt to cope with the limitations of the major currents in psychology that were prevalent in the first decades of the 20th century (behaviorism, introspective psychology and psychoanalysis). The theory was grounded on Lev Vygotsky's cultural-historical psychology,[134] and criticized the fact that scientists conceptualized psychological phenomena as autonomous and self-existent, leaving out the analysis of the relation between the subject and the surrounding social environment. Leontiev suggested that through the category of *activity* we can override the dichotomy between the external objective world and internal psychic phenomena, as suggested in the behaviorist stimulus–response pattern of early psychology. If the question now was the relation and connection between external and internal activity, in Activity Theory "it is *external activity that unlocks the circle of internal mental processes*, that opens it up to the objective world."[135]

The central aim of activity theory, in the context of psychology and social science, is to understand human beings on the basis of the analysis, genesis, structure and processes of their *activities*.[136] Activity is conceived as an intentional interaction of human beings with the material world and within the system of relations of a society, which achieves mutual transformations in the process. It is a connecting link that mediates between subject and object, an "intertraffic between opposite poles."[137] It is "[a] unit of life mediated by mental reflection, by an *image*, whose real function is to orientate the subject in the objective world."[138] The theory emphasizes the centrality of activity, because no property of the subject or the object exists before or beyond this activity. Although the circular nature of the processes that affect our interactions with the environment cannot be denied, says Leontiev, any mental reflection of the objective world, with its specific structure and material properties, is not merely due to these external influences, but generated by our practical contact with it, i.e. the subject's external activity.[139]

Activity is the basic unit of analysis and understanding of *subjects* and *objects*, and cannot be achieved by focusing on the subject or the object separately. It cannot exist outside its social conditions, which carry the aims and motives that produce it.[140] Unlike actor–network theory and distributed cognition theory, in activity theory, interaction between subject and object is not defined by symmetrical relations of system components. In activity theory any activity is the activity of the corporeal life of the material *subject*; subjects live in the world and have needs that they can fulfill only by acting on that world.

In activity theory the object has the property of a *motive*, because it motivates and directs the subject to fulfill a certain need, thus realizing an active relationship with the object. The object appears as both in its independent existence and as the mental image of the subject, effected only by the activity of that subject. Hence an "objectless activity" makes no sense according to Leontiev, and although an activity may appear to be objectless, if it had to be studied it should be analyzed by discovering its object. When the need that motivates the subject is combined with the object, then an activity emerges (which is directed towards the object). Thus activity is a unit of the interaction between subject and object, which is defined by the motive of the subject: the objectification of need. Activity theorists argue that an active intentional human subject participates in an activity that is oriented to a specific purpose and mediated by tools and technologies, which interconnect that subject with the world. Mediating artifacts can, in turn, be made and changed by those who use them – they may be products of our activity. As we shall see in the next section, this concept of artifact mediation, which is central to the whole theory, is also linked to phenomenological and post-phenomenological accounts of human–technology relations.

Activity theory is distinguished from actor–network theory and distributed cognition theory, because its proponents suggest that our relationship with the material culture, tools and machines is asymmetrical. Distributed cognition theory and actor–network theory are concerned with large-scale actants, with flat homogeneous nodes (human and non-human), whereas activity, in activity theory, is directed by subjects and their intentions.[141] Victor Kaptelinin and Bonnie Nardi, analyzing the ideas, the history, the principles and the relationship of activity theory to the rest of post-cognitivist theories and their application to human–computer interaction, argue for the primary role of individuals in their relationship with the technological world:

> We believe that analysis of human embeddedness in the world, of communication and collaboration within the social context, is of fundamental importance for understanding our relationship to technology. Human beings cannot be reduced to parts of larger-scale entities determining the nature and the meaning of the parts.[142]

Kaptelinin and Nardi suggest that, although humans conform to typical aspects of collective activities, such as rules, roles and routines, they do not follow them always, and sometimes they may change them. Although, in the context of activity theory, both individual subjects and artifacts are parts of the system, only the former have their own needs and motives:

> What we propose is that changes in both the subject and the system cannot be fully explained at the system level. The individual subject develops through mediated activity in observable, measurable ways (e.g., the zone of proximal development). Systems change under the influence of observable individual actions. Deleting the individual subject deletes important sources of change, fissures that lead to transformations, and responses to fissures.[143]

Although activity theory rejects the symmetry between humans and technological things, which is advocated by actor–network theory, it does not deny the agency of technological objects. As we have already suggested, the belief that agency belongs only to people is rather naïve, given the developments in artificial life research, which produces entities with apparent social and emotional attributes. This addresses an important difference between traditional tools and intelligent objects, environments and machines. Nevertheless, unlike actor–network theorists, who see agency as equally distributed in people and things, the proponents of activity theory argue that there are different *forms* of agency. Absolute symmetry does not exist, because people – unlike intelligent machines – have *intentions* and *creativity*.[144] They can develop their intentions based on their needs and fulfill them by acting on other entities, both human and non-human. The need for action, which activity theorists talk about, combines biological and cultural needs; thus, all living beings have biological needs (survival and reproduction), but only people have cultural needs as well. Social organizations, the world trade organization, the United Nations, are all actors that have both the need to act and agency. Yet, their motives are not biological but only cultural. Machines have the ability to act (acting in specific ways in the social environment), but they do not have the need to act; they can implement the intentions and fulfill the needs of other people. Therefore, whether an entity or a thing has agency cannot be simply answered by a yes or no, because there are different levels to this attribute:

> Thus agency should not be considered a monolithic property that is either present or absent in any given case. [...] [W]hat we propose is a combination of (1) a strict subject–object dichotomy (and resulting asymmetry) and (2) the notion of *levels of agency*, an understanding of agency as a dimension rather than a binary attribute.[145]

Kaptelinin and Nardi propose three types of agency: need-based agency, based on biological or cultural needs; delegated agency, namely the implementation of the intentions of someone else (for instance, the agency of the industrial assembly line is delegated by the industry manager); and conditional agency, which does not presuppose intentions (for instance, although a tsunami causes disaster, it has no intention to do so).[146]

Thus, the theory as outlined by Kaptelinin and Nardi retains asymmetry, but this does not mean that activity belongs to an isolated subject acting on a passive environment. The belief that things have some kind of agency can be traced in all cultures historically, including primitive animism and contemporary anthropomorphism of electronic devices. Yet, as explained in Chapter 3, anthropomorphism is people's innate tendency to attribute human mental and psychological properties to non-living things. Although Kaptelinin and Nardi understand that attributing intentions and desires to living objects (such as the applications of artificial life) are mostly projections of human desires and fears, at the same time, they suggest that the field of A-Life can be a possible future source of intentional artifacts.[147]

Embodied Interaction, Phenomenology and Post-phenomenology

As discussed in Chapter 2, embodied cognitive science scholars argue that physical embodiment is a necessary condition for mental processes, by contrast to the Cartesian dualist view of classical cognitive science and AI.[148] The emphasis is on the role of sensory-motor engagement of organisms with their physical environment in cognitive processes. Paul Dourish's theory of embodied interaction discussed in his book *Where the Action is: The Foundations of Embodied Interaction*, shares with embodied cognitive science the view that embodiment, the physical embeddedness of an active participatory subject in the real world, is necessary for the creation, manipulation and sharing of meaning with the world.[149] As Dourish argues: "embodiment is the property of our engagement with the world that allows us to make it meaningful."[150] Like activity theory, the actions of subjects in embodied interaction theory are also essential for understanding the world and its meanings. But unlike radical advocates of embodied cognition, who take a full nonrepresentational stance towards cognition,[151] Dourish, who mostly focuses on the design of software systems (already representational mediums), questions but does not reject the role of representations in interaction design. To this end, he lays out a theory to propose general principles and directions for the design of interactive environments and artifacts, considering the centrality of embodiment in human–computer interaction:

> What I am claiming for "embodied interaction" is not simply that it is a form of interaction that is embodied, but rather that it is an approach to the design and analysis of interaction that takes embodiment to be central to, even constitutive of, the whole phenomenon. This is certainly a departure from traditional approaches to HCI design.[152]

Dourish grounds his theory on the philosophical tradition of phenomenology, which, as he argues, provides a point of departure for a fundamental understanding of and a foundational approach to embodied interaction. The phenomenological concept of embodiment refers to the common way by which we confront and experience the social and physical reality of the everyday world. Phenomenologists argue that our experiences depend on our interactions with the world, and the possibilities for actions that are provided by this world (either through physical form or through socially constructed meanings). Husserl's emphasis on the phenomena of experience, Heidegger's reconstruction of Husserl's phenomenology through the concept of being as being-in-world, Schutz's extension of phenomenology to problems of social interaction, and finally Merleau-Ponty's addressing of the role of the body in perception, explore the relations between embodied action and meaning, placing the source of meaning of things already in the world where humans act and which, in turn, acts on us.[153]

This meaning reveals itself through actions in the world, an idea that significantly echoes James Jerome Gibson's ecological theory of perception in environmental psychology and his theory of affordances.[154] Gibson's theory of affordances was brought into the field of design and HCI by Donald Norman in the late 1980s. In his book *The Design of Everyday Things*, Norman used the term to refer to the fundamental properties of things (perceived and actual) that determine how they will be used.[155] He made the distinction between real affordances (the ones discussed by Gibson which are not necessarily known by users) and *perceived* affordances, namely what the user perceives the artifact can do. This concept is useful for designers who want to make the functionality and internal structure of artifacts (either physical or digital) visually perceivable through meaningful accessible representations and understandable conceptual models.[156] Later, however, Norman dropped the concept to replace it with a more viable term, the *signifier* (the perceivable part of the affordance), and a more specific term, the *social signifier*. This is an indicator, a perceivable clue that provides meaningful information about the physical or social world, informing users about what an artifact can be used for.[157]

Dourish's main interest is to examine the close ties between physical interaction within computational environments and embodied interaction founded on phenomenology. He also wants to show that the two main areas of design he calls "tangible" and "social computing" "draw on the same sets of skills and abilities, and are aspects of the same research program."[158] Both tangible computing (that will be examined further in Chapter 5) and social computing, exploit our familiarity with the everyday world starting from the view that, as embodied participatory subjects, we act in real-world settings (including embodied physical artifacts participating in the activities of its subjects), either though physical affordances and configurations (in the case of tangible computing) or through socially constructed meanings (in the case of social computing). By proposing the foundation of embodied interaction within the phenomenological tradition of embodiment, Dourish aspires to explore the close ties of tangible and social computing on the one hand, and on the other, to show that his approach can inform and support the design analysis and evaluation of interactive systems from the perspective of embodiment.[159]

In this case Dourish's approach, which understands *embodied interaction* as "the creation, manipulation, and sharing of meaning through engaged interaction with artifacts,"[160] has close affinities with Peter-Paul Verbeek's view. In his book *What Things Do* (2005), Verbeek proposes a philosophical framework, through a close reading of Heidegger (especially his early work), Don Ihde and other thinkers, to suggest that meaning emerges in direct engagement with objects. While Dourish looks into the potential applications of his theory in human–computer interaction (involving both physical and symbolic representations in software systems), Verbeek is interested in the physical articulation of devices and technological artifacts that are not necessarily computational or interactive.

Verbeek examines how technology mediates human–world relations with respect to perception and action, by taking a "post-phenomenological" approach to the philosophy of technology – following the work of philosopher Don Ihde[161] – through a thorough and nuanced critique of the classical philosophy of technology, represented, in Verbeek's book, by Karl Jaspers and Martin Heidegger.[162] Verbeek's main critique of these thinkers' view of the role of technology in mediating human–world relations is that they follow a "transcendentalist" and abstract style of reasoning, reducing technology to its conditions of possibility, its origins, leaving no room for different kinds of descriptions of different kinds of technologies. Thus, they fail to come to terms with the variety of mediations provided by specific concrete technological artifacts and practices. These thinkers' views have also been challenged by empirical studies, which brought to light the differentiated local and contextual nature of different kinds of technologies.[163] But Heidegger's early work, Verbeek argues, provides an investigation into the role actual technologies and tools play in constituting networks of meaning between humans and their world. In this work, Heidegger provides the foundations for a different approach to the philosophy of technology, in which technologies contribute to shaping a world by mediating human action. Instead of focusing on technology's conditions of possibility, this approach examines the world of concrete artifacts and empirical investigations.[164] Yet, Verbeek draws not only on the phenomenological tradition to outline his approach to technology, but also on the post-phenomenological philosophy of Don Ihde, while using insights from the work of Albert Borgmann and Bruno Latour.

Ihde puts forth a hermeneutical perspective of technological mediation, analyzing the different relations between humans and technologies in experiential terms. He proposes three different ways in which human technology relations unfold: *mediated relations*, in which our experience of reality is transformed – either strengthened or weakened – via artifacts, which can be either embodied (glasses), or hermeneutic (thermometer); *relations of alterity*, when humans relate to the artifacts themselves (think of automatic train tickets or automatic toys); and *background relations*, when artifacts shape our relations with the world, although they remain in the background (refrigerators and thermostats).[165] What is important in Ihde's *mediated relations* is that mediation does not simply take place in-between subject and object, but rather cuts across them, co-shaping and mutually constituting subject and object, humans and the world they experience.[166] But while Ihde's philosophy of mediation addresses transformations of experience, i.e. processes of amplifying or reducing perception of the world, Verbeek, following Bruno Latour's ANT, which overrides the instrumentalist view of technology, is concerned with how artifacts invite or inhibit actions, that is, mediations of action. Like artifacts in ANT, which make possible "programs of action," artifacts in Verbeek's post-phenomenological approach do act through mediations and delegations, yet this approach does not reject the gap between object and subject, as is the case with ANT, but rather attempts to bridge it, to override the dichotomy by uncovering the mutual engagements (as traditional phenomenology would have it).[167]

This technologically mediated intentionality of artifacts in Verbeek's approach follows Ihde's so-called "technological intentionality"; technologies have an implicit trajectory, promoting, but not determining, a specific way of use.[168] Of course this does not mean that artifacts have intentions the way humans do; Verbeek seems to distance himself from Latour's concept of symmetry with respect to artifacts' moral status. Latour's ANT may share with phenomenology the wish to override ontological dichotomies, but for Latour there is no gap to overcome because objects and subjects, in the ANT framework, are only products of the relations of actants in hybrid networks. Thus, while artifacts and technologies in actor–network theory mediate symmetrical relations between actants, in Verbeek's post-phenomenological perspective, artifacts are considered to be mediators of human–world relations, pointing to a mutual constitution of humans and things, actively co-shaping the way humans act in and perceive the world.[169] As we will see in the next chapter, this is a useful conceptualization about the relations between human users and their technologies in domestic environments.

Yet, in his further elaboration of the idea that technological artifacts co-shape human actions and experience, Verbeek looks into and further extends the idea of *engagement* with technological devices proposed by Albert Borgmann. He emphasizes the importance of involving users not only with the artifacts' functionality and meaning (through interfaces, buttons, signs and symbols), but also with their material machinery through sensory and bodily engagement (relations of alterity rather than embodied, where the artifact withdraws in the background, foregrounding the function and meanings that it transmits).[170] Similarly, Dourish links embodied action through natural and tangible interactions with meaning, and considers embodiment to be "a participative status, a way of being, rather than a physical property."[171] Dourish observes that in interactive systems, representations of either entities or events, symbolic or iconic, are themselves artifacts capable of being acted upon and manipulated or rearranged. On the other hand, Verbeek is critical of indirect interaction through symbolic interfaces and signs. He uses Heidegger's early insights to account for artifacts that invite participation and direct engagement with their *material* function.

Challenging Heidegger's idea of tools as either "ready-to-hand" or "present-at-hand," withdrawing from view or calling attention to themselves, Verbeek suggests that artifacts and technologies can be part of a continuum; they may promote neither full attention and involvement nor withdrawal. They can mediate action without fading into the background, and they may demand engagement by focal involvement or alterity relations (in Ihde's terms) to be useable.[172] However, for Verbeek this mediation happens through the *material* utility of artifacts – their primary function – and not through their socio-cultural utility – their secondary function. In other words, the relations that are mediated take place on the basis of the product's functionality; things mediate relations, shaping human actions and experiences, through their functioning as material objects.[173]

Thus, Verbeek is interested in the ways people physically interact with the material artifact *itself*, its mechanisms and physical properties. Apart from being a "ready-to-hand" tool, in Heidegger's terms, the artifact can be the focus of our engagement and invite participation and material involvement: "If products are to be designed to encourage human attachment, it is necessary to design them so that humans deal with the products *themselves* and not only with what they do or signify."[174] Dourish on the other hand, following Heidegger's conceptualization of artifacts as "ready-to-hand," emphasizes that, when technologies are used, they "disappear" from our view and immediate attention.[175] Thus, while Verbeek considers the design of physical artifacts that afford direct involvement and participation or, in Ihde's terms, relations of alterity (rather than "embodied"), Dourish is interested in tangible interfaces in HCI, without separating the physical representations from the symbolic realm (the symbolic representations) that they are attached to.[176]

Conclusions

We have discussed different approaches to the problem of subject–object interaction, focusing in particular on human–machine interaction. Using human conversation metaphors for the design of computationally augmented architecture does not seem to be sufficient, given the asymmetry of human–machine interaction that the theories of asymmetrical relations support. As Norman argues, smart machines should not try to understand the intentions and motives of people with whom they interact or predict their next actions, because this would be an impossible task; the unpredictable and incomprehensible actions of the machine, combined with the inability of the machine to understand the user, would drive interaction to confusion. Because the common ground required for the human–machine dialogue to be feasible and effective does not exist, it is preferable for the machine to behave predictably and let human users respond accordingly.[177]

Since interaction is asymmetrical, the design of interfaces, as Suchman argues, should draw on alternative ideas about interaction rather than on the simulation of human communication.[178] The key is not to create "smarter" and technologically better systems, Norman argues, but a different perception of the concept of interaction. Therefore, we need to design systems that avoid the problem of the lack of common context, that suggest rather than demand; systems "…that allow people to understand and choose rather than confronting them with unintelligible actions…."[179] In his later book *Living with Complexity* (2011), Norman addressed the idea of "sociable machines," namely machines that can understand their users' point of view, and can provide the necessary information to make their workings understandable. These machines could cope with lack of comprehension between humans and machines.[180]

At the same time, the relationship of human and machine in interaction, although asymmetrical, takes place between entities that have some level of agency. The relationship that develops is not one of master and servant, controller and controlled. The machine is not a passive object totally controlled by users, but rather an actant able to exercise its own action in turn. Agency can be a programmed mode in the software that runs an intelligent machine (a sort of embodied "motive"), and it can be acquired through experience. Thus, users, as active intentional subjects, can apply meaning onto the system by actively interacting with it, through situated and embodied exchanges.

This chapter has reviewed several theories to suggest a framework to explore the concept of interaction in augmented environments. In the next chapter we will look at specific criteria for the application of these theoretical ideas in practice, in order to propose the concept of a functionally open-ended, interactive and user-driven environment.

Notes

1 Fox and Kemp, *Interactive Architecture*, 13.
2 Nimish Biloria, "Introduction: Real Time Interactive Environments," in *Hyperbody*, 368.
3 Nimish Biloria, "Introduction: Hyperbody. Engineering an Innovative Architectural Future," in *Hyperbody*, 176–177. Also Christian Friedrich, "Immediate Architecture," in *Hyperbody*, 225–260.
4 Thomasz Jaskiewicz, "(In)formed Complexity. Approaching Interactive Architecture as a Complex Adaptive System," in *Hyperbody*, 184–185.
5 Kas Oosterhuis, "Introduction / Editorial," in *iA#1 – Interactive Architecture*, eds. Kas Oosterhuis and Xin Xia (Heijningen: Jap Sam Books, 2007).
6 Usman Haque, "Distinguishing Concepts: Lexicons of Interactive Art and Architecture," *Architectural Design* (4dsocial: Interactive Design Environments) 77, 4 (2007): 26.
7 Daan Roosegaarde, "Liquid Constructions," in *GameSetandMatch II: On Computer Games, Advanced Geometries and Digital Technologies*, eds. Kas Oosterhuis and Lukas

Feireiss (Rotterdam: Episode Publishers, 2006), 167. See also the webpage of the project at www.studioroosegaarde.net/project/4d-pixel.
8 See the video at https://www.youtube.com/watch?v=PVz2LIxrdKc.
9 Hosale and Kievid, "InteractiveWall," 485.
10 Ibid., 484–496.
11 Oungrinis and Liapi, "Spatial Elements Imbued with Cognition," 424.
12 See the project webpage at http://www.tielabtuc.com/#/spirit-ghost.
13 Oungrinis and Liapi, "Spatial Elements Imbued with Cognition," 431–435.
14 Hugh Dubberly and Paul Pangaro, "On Modelling: What is Conversation and how can we Design for it?," *Interactions – The Waste Manifesto* 16, 4 (July/August 2009): 22–28.
15 The *Interactive Wall* was an orthogonal surface that deformed successively as a human figure walked by. These deformations were supposedly triggered by the movements of the figure. The idea was to explore interactivity between human beings and architecture. Although this project recalled dECOi's Aegis Hyposurface mentioned in Chapter 3, it aimed to emphasize the possibility of "conversation" between humans and inanimate structures by exchanging physical motions.
16 Usman Haque, "The Architectural relevance of Gordon Pask," *Architectural Design (4d Social: Interactive Design Environments)* 77, 4 (2007): 27.
17 Gere, *Digital Culture*, 66–69.
18 For instance Vannevar Bush, in his article "As we may think" (*Atlantic Monthly*, 1945), described a theoretical machine which he termed "memex," that would be able to enhance human memory, allowing the user to store and retrieve data and files, which were connected through associations. This property of connectivity is similar to what we currently call hypertext. See *Internet Pioneers: Vannevar Bush*, ibiblio, accessed February 8, 2008, http://www.ibiblio.org/pioneers/bush.html.
19 As, for instance, is suggested in Horst Oberquelle, Ingbert Kupka, and Susanne Maass, "A view of Human–machine Communication and Cooperation," *International Journal of Man-Machine Studies* 19, 4 (1983): 313.
20 "Associative indexing" would allow users to record trails of information for other users to follow; it is a precursor of the properties of contemporary hypermedia and multimedia.
21 Lister et al., *New Media*, 40–41.
22 Pierre Lévy, *Cyberculture*, trans. Robert Bononno (Minneapolis: University of Minnesota Press, 2001), 61–65.
23 Lister et al., *New Media*, 20–21.
24 Gere, *Digital Culture*, 213.
25 Donald A. Norman, *Things that Make us Smart: Defending Human Attributes in the Age of the Machine* (Boston MA: Addison-Wesley Publishing, 1993), 243–249.
26 Gere, *Digital Culture*, 213–214.
27 Rowan Wilken, "Calculated Uncertainty: Computers, Chance Encounters, and 'Community' in the Work of Cedric Price," *Transformations: Accidental Environments* 14 (2007), accessed August 13, 2014, http://transformationsjournal.org/journal/issue_14/article_04.shtml.
28 As pointed out in Chapter 1, Gere mentions the role of other similar and related fields, such as Information Theory, General Systems Theory, Structuralism, and Artificial Intelligence, but sees them as part of a wider cybernetic culture, a way of thinking (involving developments in science, technology and art), which was the direct predecessor of our current digital culture.
29 Gere's analysis also relates these developments to the idea of the "open work" discussed by Umberto Eco in an essay where he reflects on incompleteness and indeterminacy.
30 Gere, *Digital Culture*, 81–86.
31 Mark Weiser, Rich Gold, and John Seely Brown, "Origins of Ubiquitous Computing Research at PARC in the Late 1980s," *IBM Systems Journal* 38, 4 (1999): 693–696.

32 Harvey Sacks, *Lectures on Conversation*, vol. 1–2 (Oxford: Blackwell, 1995).
33 Lucy Suchman, *Human–machine Reconfigurations: Plans and Situated Actions*, 87.
34 For instance, see Harvey Sacks, Emanuel A. Schegloff, and Gail Jefferson, "A Simplest Systematics for the Organization of Turn Taking for Conversation," in *Studies in the Organization of Conversational Interaction*, ed. Jim Schenkein (New York: Academic Press, 1978), 7–55.
35 Suchman, *Human–machine Reconfigurations*, 89.
36 Ibid., 107.
37 Ibid., 127.
38 Lucy Suchman, *Plans and Situated Actions: The Problem of Human Machine Communication* (Cambridge: Cambridge University Press, 1987).
39 Suchman, *Human–machine Reconfigurations*, 10–12.
40 Ibid., 168.
41 Ibid., 23.
42 The idea of the "common ground" was expounded by Herbert Clark and his colleagues as a mutually established background, necessary for the development of common understandings in linguistic communication. See Herbert H. Clark, *Using Language* (Cambridge: Cambridge University Press, 1996).
43 Donald A. Norman, *The Design of Future Things* (New York: Basic Books, 2007), 9.
44 Ibid., 14–15.
45 Ibid., 52.
46 Hugh Dubberly and Paul Pangaro, "On Modelling: What is Conversation and how can we Design for it?," 22–28.
47 Gordon Pask, "The Limits of Togetherness," in *Information Processing 80: Congress Proceedings*, ed. Simon H. Lavington (Amsterdam: North-Holland Publishing Company, 1980), 1003.
48 Dubberly and Pangaro, "On Modelling."
49 Gordon Pask, "A comment, a case history and a plan," in *Cybernetics, Art and Ideas*, ed. Jasia Reichardt (London: Studio Vista, 1971), 76.
50 Ibid., 88.
51 Ibid., 89–99. Also see: Gordon Pask, "The Colloquy of Mobiles," in *Cybernetic Serendipity: the computer and the arts*, ed. Jasia Reichardt (London/New York: Studio International, 1968), 34–35.
52 Pask, "A comment, a case history and a plan," 91.
53 Pickering, *The Cybernetic Brain*, 360.
54 Ibid., 319. This idea comes from Pickering's earlier work and his book *The Mangle of Practice* (1995). In this book, he studied scientific practice to propose a shift from the traditional representational idiom of science studies – science as a way to represent the world through mapping, and the production of articulated knowledge – to the performative idiom. The latter is concerned with ontological, rather than epistemological analyses, in the production of scientific knowledge, and the emergent interplay between human and material (or non-human) agency. Pickering's posthumanist or non-dualist analysis of scientific practice pointed to the "mangling," i.e. the mutual temporal coupling of people and things in an exchange of agency that he called "dance of agency." See Andrew Pickering, *The Mangle of Practice: Time, Agency, and Science* (Chicago: University of Chicago Press, 1995).
55 Pask, "A comment, a case history and a plan," 80.
56 Pickering, *The Cybernetic Brain*, 319.
57 Haque, "The Architectural relevance of Gordon Pask," 57.
58 Pickering, *The Cybernetic Brain*, 319.
59 Ruairi Glynn, "Conversational Environments Revisited" (paper presented at the 19th European Meeting on Cybernetics and Systems Research, Vienna, 25–28 March, 2008).
60 Ruairi Glynn, "Performative Ecologies (2008–10)," *Ruairi Glynn*, accessed February 2, 2013, http://www.ruairiglynn.co.uk/portfolio/performative-ecologies.

61 Ruairi Glynn, "Performative Ecologies," *Ruairi Glynn* video, 3:26, accessed February 2, 2013, http://www.ruairiglynn.co.uk/portfolio/performative-ecologies.
62 Glynn, "Conversational Environments Revisited."
63 This is something that Glynn aspires to, because mistakes and the resolution of misunderstandings are common features of human face-to-face communication.
64 Glynn, "Conversational Environments Revisited."
65 Glynn, "Performative Ecologies."
66 Glanville, "An Intelligent Architecture," 15–16.
67 See the website of ADA: The Intelligent Space at http://ada.ini.ethz.ch.
68 Eng et al., "Ada – Intelligent Space."
69 Before entering the space, the visitors had to move through a "conditioning tunnel," a corridor with interactive stations, which would introduce them to ADA's elements. Following that, in the "voyeur area," they could watch the group of people inside the room as it interacted with ADA. After entering the main space and interacting with ADA, they would enter the "brainarium," a control room where they could watch projections of its internal states in real time, relating them with the actions of the group that interacted with ADA at that specific moment.
70 Kynan Eng, Rodney Douglas, and Paul Verschure, "An Interactive Space That Learns to Influence Human Behavior," *IEEE Transactions on Systems, Man, And Cybernetics – Part A: Systems and Humans* 35, 1 (2005), 68.
71 The results of the calculation of the degree of "happiness" would lead to the choice of the most suitable behavior, which would appear on three levels. On the higher level, a neuronal configuration pattern would activate and exclude lower behaviors. Highest was the "winner-take-all" mode, followed by the "multiple-winner" mode, and on the lower level was the free behavior mode.
72 Kynan Eng, "ADA: Buildings as Organisms."
73 Cyborgs as technological beings appear early on in science fiction. In the 1940s, science fiction stories, such as C. L. Moore's *No Woman Born* and H. Kuttner's *Camouflage,* described actual cyborgs. See Katherine Hayles, "Life Cycle of Cyborgs," in *The Cyborg Handbook,* ed. Chris Hables-Gray (London/New York: Routledge, 1995), 322.
74 See Manfred Clynes and Nathan Kline, "Cyborgs and Space," in *The Cyborg Handbook,* 29–33.
75 Chris Hables-Gray, Steven Mentor, and Heidi Figueroa-Sarriera, "Cyborgology: Constructing the Knowledge of Cybernetic Organisms," in *The Cyborg Handbook,* 3.
76 This text was published in 1985 in the political theory journal *Socialist Review* under the title "A manifesto for Cyborgs: Science, Technology and Socialist-Feminism in the 1980s." In 1991 it was re-published in its final form under the title "A Cyborg Manifesto: Science, Technology, and Socialist-Feminism in the Late Twentieth Century" in Donna J. Haraway's book *Simians, Cyborgs and Women: The Reinvention of Nature* (New York: Routledge, 1991).
77 Zoë Sofoulis, "Cyberquake: Haraway's Manifesto," in *Prefiguring Cyberculture* ed. Darren Tofts, Annemarie Jonson, and Alessio Cavallaro, (Cambridge, MA/Sydney: MIT Press/Power Publications, 2002), 91.
78 Haraway, "A Cyborg Manifesto," 150.
79 Ibid. Haraway critiqued the feminist views on science and technology that only inverted western dualisms. These views prioritized the oppressed "feminine" poles (the body and matter over mind and spirit, emotion over reason, nature over technology etc.). For Haraway, in the era of biotechnology and genetic engineering, these poles were as obsolete as the "masculine" ones, which the feminist views tried to oppose. She argued that the aims of feminist politics and technological empowerment could be best served by myths and metaphors more appropriate in the era of information – not in the form of a pure natural organism, but in the form of a composite figure, such as the cyborg. See Sofoulis, "Cyberquake," 87.
80 Sofoulis, "Cyberquake," 89.

81 Katherine Hayles, *How we became Posthuman: Virtual Bodies in Cybernetics, Literature, and Informatics*, (Chicago, IL: The University of Chicago Press, 1999), 4.
82 Ibid., 18–19. Hayles discussed the terms by which the historical construction of the posthuman was determined. First, the posthuman prioritized information over material immediacy, which means that embodiment in a biological substrate was considered to be a historical accident. Second, consciousness was conceptualized as the locus of human identity in western culture. Third, the body was considered to be the initial prosthetic device that people learn to control. Thus, it can be extended with and replaced by other devices. Fourth, the concept of the posthuman implied that the human being was capable of close articulation with intelligent machines. In this perspective there were no essential differences between bodily existence and electronic simulation, cybernetic mechanism and biological organism, robotic teleology and human goals.
83 Ibid., 4.
84 For instance, see Pat Cadigan's novel *Synners* (New York: Bantam, 1991). Also see Mark Dery, *Escape Velocity: Cyberculture at the End of the Century* (New York: Grove Press, 1996), 252–256.
85 Mark Poster, "High-Tech Frankenstein, or Heidegger Meets Stelarc," in *The Cyborg Experiments: The Extensions of the Body in the Media Age*, ed. Joanna Zylinska (London: Continuum, 2002), 29.
86 The idea that technology can be seen as a prosthetic extension of the human body was elaborated by McLuhan in *The Gutenberg Galaxy* and *Understanding Media: The Extensions of Man*. But it was reframed as an opportunity for urban planning when McLuhan attended the 1963 meeting that Konstantinos Doxiadis organized on an eight-day boat trip around the Greek islands. See Mark Wigley, "Network Fever," *Grey Room* 4, (Summer 2001): 82–122.
87 Gregory Bateson, *Steps to an Ecology of Mind* (New York: Ballantine Books, 1972).
88 William Mitchell, *Me++: The Cyborg Self and the Networked City* (Cambridge, MA/London: MIT Press, 2003), 7, 19.
89 Ibid., 39.
90 Lister et al., *New Media*, 351.
91 Ibid., 371.
92 This historically located thinking is related to the Marxist critic on capitalism, according to which human workers stand on one side of the production process of the autonomous capitalist machine and, for the first time in history, man is a component of mechanical systems, instead of being the primary actor. See ibid., 307.
93 Stelarc, "Obsolete Bodies," *Stelarc*, accessed March 2, 2014, http://stelarc.org/?catID=20317.
94 Stelarc's Third Hand experiments, with amplified body signals and sounds, included projects such as *Fractal Flesh, Ping Body, Parasite, Amplified Body and Split Body*. See Stelarc, "Fractal Flesh," *Stelarc*, accessed March 2, 2014, http://stelarc.org/?catID=20317.
95 John Appleby, "Planned Obsolescence: Flying into the Future with Stelarc," in *The Cyborg Experiments* ed. Joanna Zylinska, (London: Continuum, 2002), 110.
96 The prosthetic hand could move independently by means of electromagnetic signals transmitted by Stelarc's abdominal muscles. At the same time, his real left arm was distantly controlled by two muscular stimulation systems (electrodes placed on his flex muscles and his biceps activated physical movements of his fingers, his wrist and his whole arm). See Stelarc, "Third Hand," *Stelarc*, accessed March 2, 2014, http://stelarc.org/?catID=20265.
97 Appleby, "Planned Obsolescence," 103.
98 Poster, "High-Tech Frankenstein," 29.
99 Andy Clark, *Natural-Born Cyborgs: Minds, Technologies, and the Future of Human Intelligence* (Oxford: Oxford University Press, 2003).

100 Since 1998, Professor Kevin Warwick has examined what happens when humans and computers merge, by implanting silicon chips into his body. Unlike Stelarc's artistic performances, Warwick's experiments (which he conducted with the assistance of neurosurgeons) had foremost scientific aims: the implants could carry and transmit information about the carrier's medical records for future reference, and they could be employed to control the motion of wheelchairs or artificial hands for the handicapped. In 1998, Warwick surgically implanted a silicon chip transponder in his forearm, to test whether he could operate environmental appliances and devices (like doors, lights, and heaters) by moving around through the rooms without physical motions. In a later experiment, the Project Cyborg 2.0 (2002), he tested the possibility of exchanging signals between his body and the computer. He used implant technology (a microelectrode array within a guiding tube implanted into the median nerve fibers above his left wrist) to measure the nerve signals transmitted through his arm. He also examined whether he could exchange artificial sensations, thought, movement and emotion signals, with another person who carried a similar implant. See Kevin Warwick's webpage at http://www.kevinwarwick.com and the webpages of the experiments at http://www.kevinwarwick.com/Cyborg1.htm, http://www.kevinwarwick.com/Cyborg2.htm. Also see his book *I, Cyborg* (Champaign, IL: University of Illinois Press, 2004).
101 Sofoulis, "Cyberquake," 88–89.
102 Ibid., 101.
103 Suchman, *Human–machine Reconfigurations*, 283.
104 Ibid., 275.
105 Ibid., 261.
106 *TGarden* was an audio-visual installation where participants could explore such interconnections with the environment, wearing sensor-enhanced costumes able to track their movement in space. They could also control and interact with sound and visual effects through synthesizer media. Nothing was pre-choreographed, since participants were called to gradually learn to interact in such a way as to be able to compose their own audiovisual events. See the project website at f0.am/tgarden. Also see Maja Kuzmanovic, "Formalising Operational Adaptive Methodologies or Growing Stories within Stories," *AHDS Guides to Good Practice, A Guide to Good Practice in Collaborative Working Methods and New Media Tools Creation*, accessed June 12, 2014, http://www.ahds.ac.uk/creating/guides/new-media-tools/kuzmanovic.htm.
107 Unlike other Stelarc performances, which used hardware to augment the body, the *Prosthetic Head* was a computer-generated three-dimensional head that resembled the artist, designed to problematize questions of agency and embodiment. It worked as a conversational avatar, able to speak and perform non-verbal communication cues (through facial expressions). Using an expanding database and embedded algorithms to initiate and engage in conversation, it could interact with people who could use a keyboard to write comments and questions. See the project website at http://stelarc.org/?catID=20241.
108 Suchman, *Human–machine Reconfigurations*, 281.
109 Ibid., 285–286.
110 Bruno Latour, *Science in Action* (Cambridge, MA: Harvard University Press, 1987).
111 Most of the relevant bibliography on ANT can be found at "The Actor Network Resource" website at: http://www.lancaster.ac.uk/fass/centres/css/ant/antres.htm.
112 The actant is one of the central concepts of actor–network theory and was introduced from semiotics. It refers to material entities, human persons or groups which take on form, are defined, actualized and finally acquire agency only to the degree that they enter into alliance with a spokesperson that acts as their representative. This alliance is capable of withstanding forces that attempt to dissolve it. See: Bryan Pfaffenberger, "Actant," STS Wiki, accessed April 22, 2013, http://www.stswiki.org/index.php?title=Actant.

113 The non-human, in Latour's thinking, is not the "object" that takes the role of the passive thing in the subject–object dichotomy. The non-human has an active role and bypasses this distinction.
114 In his book *Reassembling the Social,* Bruno Latour distinguishes between the "sociology of the social," which reduces the social to a priori definitions of what constitutes it, and a subfield of social theory he calls "sociology of associations," in which the social depends on the results of its unstable assemblages and associations between things that are not themselves social. See Bruno Latour, *Reassembling the Social: an Introduction to Actor–network-Theory* (Oxford: Oxford University Press, 2005), 1–17.
115 Latour mentions about the name of ANT: "a name that is so awkward, so confusing, so meaningless that it deserves to be kept." See Latour, *Reassembling the Social*, 9.
116 John Law, "Networks, Relations, Cyborgs: on the Social Study of Technology," *Centre for Science Studies, Lancaster University,* accessed March 15, 2013, http://www.comp.lancs.ac.uk/sociology/papers/Law-Networks-Relations-Cyborgs.pdf.
117 Michel Callon, "Actor–network Theory: The Market Test," in *Actor Network Theory and After,* eds. John Law and John Hassard (Oxford: Blackwell, 1999), 185–186.
118 John Law, "Notes on the Theory of the Actor–network: Ordering, Strategy and Heterogeneity," *Systems Practice* 5, 4 (1992): 381.
119 In this, ANT is opposed to another relevant sociological field, the Social Construction of Technology (SCOT), which considers technology as a social construction and cultivates the distinctions between human/social and non-human/technological.
120 Latour explains: "The program of action is the set of written instructions that can be substituted by the analyst to any artifact. Now that computers exist, we are able to conceive of a text (a programming language) that is at once words and actions. How to do things with words and then turn words into things is now clear to any programmer." See Bruno Latour, "Where Are the Missing Masses? The Sociology of a Few Mundane Artifacts," in *Shaping Technology/Building Society: Studies in Sociotechnical Change,* eds. Wiebe E. Bijker and John Law (Cambridge, MA: MIT Press, 1992), 176.
121 Latour, "Where Are the Missing Masses?" 152–153.
122 The term "script" was coined by Madeleine Akrich in "The De-scription of Technological Objects," in *Shaping Technology/Building Society: Studies in Sociotechnical Change,* eds. Wiebe Bijker and John Law (Cambridge MA: MIT Press, 1992), 205–224. For further discussion and its relation to design see Kjetil Fallan, "De-scribing Design: Appropriating Script Analysis to Design History," *Design Issues* 24, 4 (2008): 61–75.
123 Latour, "Where Are the Missing Masses?," 176.
124 John Law, "Notes on the Theory of the Actor–network," 382.
125 See Yvonne Rogers, "Distributed Cognition and Communication," in *The Encyclopaedia of Language and Linguistics,* ed. Keith Brown (Oxford: Elsevier, 2006), 181–202. Also see Pierre Lévy, *Collective Intelligence: Mankind's Emerging World in Cyberspace,* trans. by Robert Bononno (New York: Plenum, 1997); Francis Heylighen, Margeret Heath, and Frank Van Overwalle, "The Emergence of Distributed Cognition: A Conceptual Framework," *Principia Cybernetica Web,* accessed April 2, 2013, http://pespmc1.vub.ac.be/papers/distr.cognitionframework.pdf; Peter Berger and Thomas Luckmann, *The Social Construction of Reality: A Treatise in the Sociology of Knowledge* (Garden City, NY: Anchor Books, 1966).
126 See James D. Hollan, Edwin Hutchins, and David Kirsh, "Distributed Cognition: Toward a new Foundation for Human–computer Interaction Research," *ACM Transactions on Computer-Human Interaction* 7, 2 (June 2000): 174–196.
127 See Andy Clark and David Chalmers, "The Extended Mind," *Analysis* 58, 1 (1998): 10–23. The extended mind theory studies the point of separation between the mind, the body and the environment. It proposes that the mind uses some objects in the external environment in such a way as to become its extensions (for instance, a notepad which records thoughts and reminders thus replacing brain memory

etc.). Here the mind is an extended cognitive system, interfacing human beings and external material objects (a coupled system), which actively participate in the formation of behavior, memory, beliefs and human thinking.
128 Edwin Hutchins, *Cognition in the Wild* (Cambridge, MA: MIT Press, 1995).
129 Hutchins explores the similarities and differences between western and Micronesian traditions of maritime navigation, in relation to the practices of knowledge "storage" that they developed. In the first tradition, the medieval astrolabe, which could determine spatial relations between celestial bodies at different latitudes, was a natural extension of the memory of the navigator and a locus of many generations of astronomical practices. A navigator who used the astrolabe was thus the mental inheritor of a series of embedded socio-cultural practices. On the contrary, a Micronesian navigator would not use such material artifacts but rather cultural artifacts, such as chants, which codified the celestial relations for sailing between specific islands. However, such cultural objects presented a similar function to the medieval astrolabe in that they embodied generations of knowledge gleaned by navigation practices.
130 Norman, *Things that make us Smart*, 146–147, 127.
131 For relevant discussion including the relationship between ANT and assemblage thinking see Martin Müller, "Assemblages and Actor–networks: Rethinking Socio-material Power, Politics and Space," *Geography Compass* 9, 1 (2015): 27–41, doi:10.1111/gec3.12192. For further critiques of ANT see David Bloor, "Anti-Latour," *Studies in History and Philosophy of Science* 30, 1 (1999): 81–112, doi:10.1016/S0039-3681(98)00038-7; Frédéric Vandenberghe, "Reconstructing Humans: A Humanist Critique of Actant-Network Theory," *Theory, Culture and Society* 19, 5–6 (2002): 55–67; Harry M. Collins and Steve Yearley, "Epistemological chicken," in *Science as practice and culture*, ed. Andrew Pickering (Chicago: University of Chicago Press, 1992), 301–326; Andrea Whittle and André Spicer, "Is actor–network theory critique?" *Organization Studies* 29, 4 (2008): 611–629, doi:10.1177/0170840607082223; Olga Amsterdamska, "Surely You're Joking, Mr Latour!" *Science, Technology, Human Values* 15, 4 (1990): 495–504.
132 Andrew Pickering, "The mangle of practice: agency and emergence in the sociology of science," *American Journal of Sociology* 99, 3 (1993): 559–589. See also Langton Winner, "Upon Opening the Black Box and Finding It Empty: Social Constructivism and the Philosophy of Technology," *Science, Technology, & Human Values* 18, 3 (1993): 362–378.
133 See Aleksei N. Leontiev, *Activity, Consciousness, and Personality*, trans. Marie J. Hall (Hillsdale: Prentice-Hall, 1978). Also see Aleksei N. Leontiev, "Activity and Consciousness," in *Philosophy in the USSR, Problems of Dialectical Materialism* (Moscow: Progress Publishers, 1977), 180–202. First published in *Voprosy filosofii* 12, (1972): 129–140. See also James Wertsch, ed., *The Concept of Activity in Soviet Psychology* (Armonk, NY: M. E. Sharpe, 1981).
134 Vygotsky was Leontiev's teacher and colleague. The main argument of Vygotsky's theory was that any analysis of the mind should include an analysis of the culturally and socially determined interaction between people and the world in which it is embedded. Many theories and ideas that developed after the Bolshevik revolution of 1917 – among them Vygotsky's – argued for the union of consciousness and activity, as well as the social nature of the mind, according to Marxist philosophy. See: Victor Kaptelinin and Bonnie Nardi, *Acting with Technology: Activity Theory and Interaction Design* (Cambridge, MA: MIT Press, 2006), 36–37.
135 Leontiev, "Activity and Consciousness," 184.
136 Leontiev differentiates between *Activity* and *Action*. Whereas the former is driven by motives, the latter is goal-oriented. For instance, if a hunter's motive is to obtain food, his goal is the trapping gear, which he has to construct by performing actions that are oriented towards that goal. Thus, a given action may realize completely

different activities, whereas the same motive may generate various goals and therefore various actions. Ibid., 185–187.
137 Ibid., 182.
138 Ibid.
139 Ibid., 183.
140 Ibid., 182.
141 It should be noted that distributed cognition theory, despite its similarity with actor–network theory on the question of symmetry, differs from the latter in that it is interested in the way objects store knowledge, thoughts and memory, implementing a wider, extended cognitive system. On the other hand, actor–network theory examines agency beyond human beings, in distributed networks of both social and material/technological components.
142 Ibid., 207.
143 Ibid., 235.
144 Ibid., 238–241.
145 Ibid., 247.
146 Ibid., 247–248.
147 Ibid., 250.
148 Of course embodied cognitive science is an interdisciplinary field of research, borrowing largely from areas such as the philosophy of mind and phenomenology, cognitive science, psychology, cognitive neuroscience and artificial intelligence (robotics). See Lakoff and Johnson, *Philosophy in the Flesh*; Clark, *Being There*; Pfeifer and Bongard, *How the Body Shapes the Way We Think*; Brooks, *Cambrian Intelligence*; Varela, Thompson, and Rosch, *The Embodied Mind*.
149 Paul Dourish, *Where the Action is: The Foundations of Embodied Interaction* (Cambridge, MA/London: MIT Press, 2001), 126.
150 Ibid., 126.
151 See for instance Clark, *Being There*.
152 Dourish, *Where the Action is,* 102.
153 Although the philosophy of embodiment follows different paths in each of the phenomenological traditions, all four share the belief that coming in contact with and understanding of the world is a direct rather than an abstract or ideal process. Ibid., 116–117.
154 Gibson argues that environmental perception is direct and not mediated by internal processes. Unlike constructivist theories of perception, as well as the computational model of perception in cognitive psychology, ecological theory favors the view that there is sufficient information in the environment for direct perception. Such information comprises the innate characteristics of visual images, texture gradients, motion parallax, shadows and occlusion, which can determine the depth, inclination, size or the distance of a surface. See James Jerome Gibson, *The Ecological Approach to Visual Perception* (Hillsdale, NJ: Lawrence Erlbaum Associates, 1986 [1979]).
155 Donald A. Norman, *The Design of Everyday Things* (New York: Doubleday/Currency, 1990), 9. Originally published as *The Psychology of Everyday Things* (New York: Basic Books, 1988).
156 Norman, *Things that make us smart*, 104–106. Also Norman, *The Design of Everyday Things*, 9–33.
157 On the idea of the "social signifier" see Donald A. Norman, *Living with complexity* (Cambridge, MA/London: MIT Press, 2011), 88–108. Also see Donald Norman, "THE WAY I SEE IT: Signifiers, not Affordances," *Interactions – Designing Games: Why and How* 15, 6 (November/December 2008): 18–19.
158 Dourish, *Where the Action is*, 17.
159 Ibid., 17–22.
160 Ibid., 126.
161 Post-phenomenology is a modified, hybrid philosophy of technology that blends classical phenomenology with pragmatism. By allowing pragmatism to move towards

phenomenology, post-phenomenology avoids subjectivism and antiscientific idealism of classical phenomenology. Conversely, post-phenomenology uses insights from variational theory, the phenomenological notions of embodiment and the lifeworld to enrich the analysis of the experimental in pragmatism. Most importantly, a post-phenomenological approach steps away from generalizations and transcendental perspectives about technology, towards "the examination of technologies in their particularities" (an idea that Ihde shares with Dutch philosopher of technology Hans Achterhuis), emphasizing the use of concrete and empirical studies of technologies. See Don Ihde, *Postphenomenology and Technoscience: The Peking University Lectures* (Albany, NY: State University of New York Press, 2009), 5–24.
162 Jasper addressed the alienation that technology has brought to human existence, from an existential phenomenological perspective, while maintaining at the same time an instrumentalist view, i.e. that technology is a neutral means driven only by human goals. This assumes that humans can reverse technology's "demonic," threatening, and independent power, to recover their mastery over it and regain their humanity. Heidegger's hermeneutical phenomenology, on the other hand, attempts to uncover the general mode through which reality becomes disclosed to us; in his later thinking, Heidegger views technology as a dominating and controlling manner of approaching reality, which is a raw material subject to manipulation.
163 Peter-Paul Verbeek, *What things do: Philosophical Reflections on Technology, Agency and Design* (University Park, PA: Pennsylvania State University Press, 2005), 5.
164 Ibid., 85. As Verbeek explains, this new philosophy "is a forward thinking, starting from the technologies themselves and asking what role they play in our culture and daily lives, instead of reducing them to the conditions of their possibility and speaking about these conditions as if we were speaking about technology itself." See ibid., 9.
165 Ihde, *Postphenomenology and Technoscience*, 42–43.
166 Verbeek, *What things do*, 130.
167 Ibid., 166.
168 Ibid., 114–115.
169 Ibid., 163–168.
170 Ibid., 228–229.
171 Dourish, *Where the Action is*, 125.
172 Verbeek, *What things do*, 194.
173 Ibid., 209.
174 Ibid., 232.
175 Dourish, *Where the Action is*, 109–110.
176 Ibid., 207.
177 Norman, *Things that Make us Smart*, 76–77.
178 Suchman, *Human–machine Reconfigurations*, 178–179, 183.
179 Norman, *Things that Make us Smart*, 55.
180 Norman, *Living with complexity*, 115.

5

TANGIBLE COMPUTING

Towards Open-Potential Environments

We have already outlined a theoretical framework to discuss the interactive aspects of computationally augmented environments. In this chapter we will look at an alternative Human–Computer Interaction perspective, which pertains to tangible computing systems and end-user driven environments. These systems allow users to program and customize their environment, by constructing their own domestic functions according to their changing needs. We will argue that these environments constitute "ecologies" of interacting artifacts, assembling heterogeneous components, human users, infrastructure and software, in "conversational" processes. Along these lines and paradigms we will then outline our proposal for the "open potential environment," a system that can be implemented by means of user-driven tangible computing systems in domestic spaces, and can demonstrate open-ended, indeterminate and multiple functionalities. This approach to design, which deals with domestic environments and infrastructures in the household that are functionally determined by users, is entangled with questions about the notion of domesticity, changing patterns of spatial and temporal organization of the household, the influence on the rituals of daily life, as well as novel understandings of spatiality.

Tangible Computing and End-user Driven IEs

Augmented Artifacts

In the context of his embodied interaction theory, Paul Dourish defines tangible computing as an approach to computing that explores "[h]ow to get the computer 'out of the way' and provide people with a much more direct – tangible – interaction experience."[1] Tangible computing applications, which range from work in computer science, as well as art and design (so-called computer-related or interaction design), assume the distribution of computational power across different devices or objects that can behave as active agents, reacting to the activities of people, while knowing their location and context. Instead of traditional computational devices, such as graphical computer interfaces and keyboards, tangible computing uses physical artifacts and their affordances in three-dimensional physical space.

There have been several examples and projects since the 1990s – from Durell Bishop's *Marble Answering Machine* and the work of the computer-related design department at the Royal College of Art directed by Gillian Crampton-Smith, to the work of the Tangible Media group at the MIT Media Lab, directed by Hiroshi Ishi.[2] Designers have studied not only how computation moves into the world and out of the desktop PC (which is the first topic that tangible computing deals with), but also how computation may manifest itself when it moves into the physical environment (the second topic), and how physical manipulation of digitally enhanced tangible artifacts could help naturalize the ways we interact with computation (the third topic), and thus make those interactions meaningful: "[t]he essence of tangible computing lies in the way in which it allows computation to be manifest for us in the everyday world; a world that is available for our interpretation, and one which is meaningful for us in the ways in which we can understand and act in it."[3]

Tangible computing objects incorporate a "digital self" – hardware and software components – that represents the artifact, in the digital space, for other artifacts.[4] Digital information is thus embedded in physical form while manifesting itself through physical changes in the domestic environment: for instance, lighting patterns, sound signals or movement of furniture and architectural elements. In tangible computing systems, there is no single point of interaction, a single device to interact with, as in traditional human–computer interaction interfaces. The same activity can be distributed and shared between multiple devices or, more specifically, it can be achieved through the coordinated use of these artifacts.[5] At the same time, interaction does not take place in a successive manner of information exchange, but in parallel, as there is no way to tell the next actions of users. Tangible computing artifacts have several properties that distinguish them from traditional objects: they possess information, instructions for the execution of their functions, and messages that they can exchange wirelessly with other artifacts (from simple programs to complex databases); they can also process this information, which appears in the digital space as data about services provided by the artifact or requested by other artifacts; they can interact with their environment, by sensing their context of use and by communicating with other artifacts; and they can respond to these stimuli using actuators. Tangible computing artifacts can collaborate. Using their embedded communication unit and the capacity to publish their services, they can be combined with other objects, as virtual building blocks of larger and more complicated systems. They can also change or disengage the digital services provided.[6]

In traditional human–computer interaction, designers fully control the functionality, use and form of their artifacts. They start from pre-specified design models to anticipate the way these artifacts will interact with their users.[7] But often users interact with and apply meaning onto augmented artifacts in surprising ways, which presents a possible alternative approach to design thinking for augmented artifacts. Meaning is constructed and communicated on many levels within a process that Dourish calls "coupling," the intentional connection between users and artifacts that arises in the course of interaction:[8] "So, again, because meaning is an aspect of use, interaction, and practice, it is something that resides primarily in the hands of the user, and not of the designer."[9] In this case, users, not designers, are the agents that actively determine how technology will respond to their constantly changing needs and how they will use artifacts to perform a certain function.

This changes the role and presence of designers in interaction design. Instead of designing and anticipating the way people will use interactive objects, designers should explore methods to allow users to appropriate the objects, to act both on and through those objects. Users should be able not only to execute the normal functions of artifacts, but also to modify their interface, to break it and appropriate it by relocating its constituent elements.[10] This relates to Verbeek's reference to Ihde's concept of "multistability" of technologies, which opposes determinist views of technology. Ihde uses this concept to address the idea that technologies cannot be separated from their use and cultural context; they are always undetermined, with partially determined trajectories, acquiring their identity only in use and through their relations to humans.[11] Yet, as Verbeek explains, this does not mean that designers cannot anticipate the mediating role of technologies, since research into the habitual use and conventions of artifacts can be taken into account.[12] But, unlike Dourish, who does not distinguish between objects and their representations, Verbeek, as discussed in Chapter 4, focuses on users' involvement with the *material* function of artifacts, de-emphasizing their visual aspects and interface. This is demonstrated, for example, in the *Ithaca* color printer, which allows users to engage actively with its material machinery through its "transparency" and "functional clarity."[13]

We address these considerations here because they can inspire the design of domestic, user-customizable augmented environments. Users, as active agents, can define the function of their environments within a prescribed range of options, services and system flexibility. The system allows them to create unanticipated functionalities, which are no longer controlled by the designer; the designer is only responsible for determining the range of possible functions. As discussed in Chapter 2, end-user driven intelligent environments, such as *PiP* (pervasive interactive programming) and e-*Gadgets*, enable users to program their environment by "building" their own virtual functions and adapt them to their indeterminate needs; they are able to construct their personal AmI applications in domestic environments. Through user-friendly interfaces, these systems can empower users to appropriate the system and its operational rules in a creative and even improvisational way.[14] In the *PiP*, users can deconstruct the system by choosing and combining different device functions, thus forming "virtual pseudo-devices" (Meta-Appliances – Applications).[15] *E-Gadgets*, which is examined below, comprises computationally augmented tangible objects that can be functionally interconnected by users to perform custom and personalized operations.

E-Gadgets

Extrovert-Gadgets (2000–2003) was a research project conducted under the Information Society Technologies, Future Emerging Technologies (IST-FET) and Disappearing Computer initiative of the EU. Its researchers wanted to create open-ended and customizable ambient intelligence environments, by embedding hardware and software components into everyday tangible objects, such as lights, switches, cups, PDAs, hi-fis, pens, keys, books, offices, TV sets or even rooms and buildings. In this way, apart from their physical affordances (form, shape and location), artifacts acquired digital capacities: a digital unit, comprising micro-processor, memory, sensors, actuators, and wireless communication. Each e-gadget performed as an independent autonomous entity, because it could operate without knowing the actions of other e-gadgets. The components were able to manage their resources locally (such as power), to function according to their inbuilt operations, and to communicate directly or indirectly with other e-gadgets to create synapses.

The project extends the notion of "component-based architecture," and adapts it to the world of tangible physical objects and spaces. Component-based architecture in software engineering consists of software components, construction modules belonging to larger systems that can publish their functionalities, connect and collaborate with other components in the system. According to the research project coordinator Achilles Kameas, what distinguished this project from other intelligent environments was that nothing was prescribed by design; the objects were heterogeneous and people were free to act as they wished.[16] For an environment to be able to adapt to changing and unstable circumstances, it must have some information about its context and its specific situations. Because these situations cannot be predetermined by the designers, the challenge was to provide the tools for users to determine, adapt and customize or even construct their own AmI applications.[17]

Each e-gadget included a board that ran the GAS-OS middleware, a wireless P2P network to communicate information about its physical properties (color, weight, shape and geometry etc.), and integrated sensors (such as light sensitive pressure pads and infrared sensors) to communicate information about its state, such as weight, lighting, temperature, distance from human users etc. A chair for instance could send pressure and weight information if a person sat on it. The middleware enabled the mechanism to facilitate interaction and cooperation between heterogeneous e-gadgets (constructed by different manufacturers); it operated as a mediator, able to find and publicize the services and properties of the artifacts, using a common vocabulary (namely their representation with terms and concepts that are common to all e-gadgets).[18] The middleware implemented the digital self of the e-gadgets, namely the plugs and their connections. The plugs were virtual entities, part of the artifact's interface, that could communicate the properties (such as information coming from one or more sensors) and the provided services (what the object can do) of each e-gadget to the users and to other e-gadgets. At the same time the plugs enabled the creation of *synapses* between the artifacts, namely local interconnections between the plugs of the e-gadgets, which were made by users according to their custom needs. When users wanted to create, modify or destroy synapses between different e-gadgets, they would use a graphic editor to process that information and suggest preferred operations.

The idea that inspired the designers of the project came from what happens in the physical world of everyday personal space. People customize their environments, especially domestic spaces, arranging the objects that they contain according to their personal preferences and activities. Users of e-gadgets can program their spaces by determining the combinations, drawing lines between plugs in the graphic editor or writing rules ("if…, then…") on the synapses. In this way e-gadgets could be regarded as "black boxes" – flexible building blocks – with no predefined capabilities, because they have an "open" operation mechanism. An implemented example connects a desk (e-Desk), which can sense a nearby chair (e-Chair), a lamp on the desk (e-Lamp) and a book (the e-book). Users can create a synapse so that when they sit on the chair and open the book on the desk, the lamp is turned on. Through the chair plug, information is transmitted to the plug of the lamp, in order to send the message to turn it on.[19] This capability of the ambient environment that comprises the desk, chair, book and lamp is determined only by users through the synapses, namely the following customized and personalized operational rules:

> *IF the chair is close to the desk*
> *AND*
> *Any book is on the desk,*
> *AND*
> *Someone is sitting on the chair,*
> *AND*
> *The book is open*
> *THEN*
> *Turn on the lamp.*[20]

In this way users create an immaterial sphere (AmI sphere) of functional relations, a distributed system of e-gadgets, the "e-GadgetWorld," which emerges as a unified entity from the different synapses and connections that are defined or modified by users. The e-GadgetWorld is a dynamic functional configuration of interconnected objects that comprise the ambient environment, able to perform an overall collective function as determined by an intentional agent (a person). This function defines the overall state of the environment that is different from the state of each individual e-gadget. In the aforementioned example, this functional configuration is a collection of synapses ordered to perform a "reading mode." Throughout this process of defining the performance mode, users may experience unexpected behaviors, if, for instance, they interconnect arbitrarily unrelated things. Anything that has an output can be connected to something that has an input, to construct a synapse. The user will then decide whether this synapse is meaningful or not.[21]

Through the graphic editor, users can create improvisational and indeterminate functional interconnections between artifacts. At the same time, intelligent agents can access the editor in order to operate the system autonomously, having previously recorded and stored user behaviors. Thus, an e-gadget can benefit from these agents; it can use them to learn from people's way of using it and store this acquired experience for future – probably autonomous – activities. The editor should be able to identify the e-gadgets and their plugs, provide information to users and allow them to choose and manage the e-gadgets, as well as their synapses and their properties. It should also support the possibility to create, activate, modify and deactivate the synapses and the e-GadgetWorlds. Also, in combination with the intelligent agent, the editor should be able to suggest synapses and applications of the system, on the basis of similar configurations that occurred previously, or to provide step-by-step consultation and instructions about the environment.[22]

The description of the e-Gadgets project seems to suggest an ambient environment of tangible objects, where the functions performed – the behaviors – are not predetermined by design, because they emerge through the indeterminate instructions provided by users, as well as the limits of the system. In fact, the capacity to create synapses between e-gadgets and the choices available to users are limited by the range of physical and digital affordances of the artifacts prescribing their use (including misuse). This limits the range of potential choices and provided services (which in the case of e-gadgets are defined by the plugs and the capabilities of the synapses). Users, for instance, can connect the seating action of the e-Chair with the automatic reduction of artificial light of the lamp (e-Lamp), but this choice is limited by the specific functions of the sensors of the chair (pressure pads) and the lamp effectors (light intensity dimmer). These limitations are, of course, related to the needs and the services required by the space within which the system is embedded.

The project does not involve physical architectural space. It focuses on the function of domestic appliances and devices, services and living conditions, which reminds us of the emphasis on domestic services and infrastructure suggested by Banham, Price and Archigram (see Chapter 1). But the ideas behind the e-Gadgets project could be expanded to enhance architectural flexibility rather than services, if, for instance, we assume that the house is composed of constituting elements, parts and components (architectural elements, walls, dividing partitions, panels, openings, doors and moveable spaces) capable of changing their configuration, state and boundaries. If these constituting parts had a board and plugs connected by synapses similar to e-gadgets, then users would be able to modify the spatial layout according to the changes of the plugs and synapses. By creating synapses, users would be able to program the location and arrangement of architectural elements or, conversely, by reconfiguring the layout of architectural elements, they could change the synapses.[23] Thus, users would be able to alter the spatial configuration on demand, according to changing needs and activities.

Another limitation is that human intervention in the e-Gadgets system takes place only through the user interface (the graphic editor); users cannot change the digital environment and the synapses through physical interaction with objects. Thus, the e-Gadgets project reflects only the first topic investigated within the research program of tangible computing: the integration of computation with physical spaces, devices and furniture that are aware of their position and neighboring artifacts. It does not reflect, however, the third topic in the tangible computing vision: programming the computational environment through physical manipulation and interaction with artifacts, rather than indirect means like graphical user interfaces.[24] Further work could be undertaken to enable actual engagement with the materiality of the physical components of the system, in the manner described by Verbeek in Chapter 4.

The idea that information can acquire physical form so that users would be able to operate the system through natural actions and interactions with everyday objects, is a fundamental approach in tangible computing. In this context, specific functionalities are linked with specific objects distributed in spaces. By adjusting and rearranging the layout of spaces and objects, users can modify and adjust the functionality of the system and hence space, adapting it to their immediate needs.[25] According to Kameas, such a perspective is theoretically feasible; in a speculative case of e-gadgets, the user could suggest rules in the synapses, so that, depending on the position and the physical arrangement of the objects, changes of their functions could also occur.[26] This seems to be the idea behind the design of the *Reconfigurable House*, which is discussed next.

Reconfigurable House

The *Reconfigurable House*, by Usman Haque and Adam Somlai-Fischer, was an experimental proposal for an open source, user-driven interactive installation exhibited at the Z33 House for Contemporary Art at Hasselt, Belgium (2008) and the ICC Intercommunication Center at Tokyo (2007). It consisted of simple sound-, light- and touch-sensitive devices, such as low-tech toys and accessible everyday components (bearing names such as "cat bricks," "radio penguins," and "mist lasers"), controlled by an Arduino open source microprocessor, which used the Processing open-source programming language (Figures 5.1 to 5.4).

FIGURE 5.1 *Reconfigurable House* built by Usman Haque and Adam Somlai-Fischer. © Courtesy of Adam Somlai-Fischer.

Users could interact either locally with the ambient environment, by changing the position of the different components, or remotely through the internet. The installation could be openly modified by its visitors: by rearranging the "hardware" (the physical components), they could affect its software and thus its operation and behavior, its sound and light events, as well as noises and mist emission.[27] Unlike other AmI environments, like the *PlaceLab* (discussed in Chapter 2), which contains built-in infrastructural apparatuses and is therefore inaccessible to users, the Reconfigurable House kept these components visible, so that users could easily rearrange them. Comparing the pre-programmed operations of conventional intelligent environments with the functional open-endedness that characterizes this project, its designers explain that the sensor and actuator components that constitute it can be interconnected in any way by the users, in order to acquire completely different and unexpected behaviors.[28] Therefore, it could be considered to be an example of *asymmetrical interaction*.

FIGURE 5.2 *Reconfigurable House* built by Usman Haque and Adam Somlai-Fischer. © Courtesy of Adam Somlai-Fischer.

FIGURE 5.3 *Reconfigurable House* built by Usman Haque and Adam Somlai-Fischer. © Courtesy of Adam Somlai-Fischer.

FIGURE 5.4 *Reconfigurable House* built by Usman Haque and Adam Somlai-Fischer. © Courtesy of Adam Somlai-Fischer.

Haque mentions that "…the Reconfigurable House also demonstrates authentic interaction: where the system not only reacts to visitors, but, at a higher level, also changes the way that its reaction is computed."[29]

But unlike the e-Gadgets project, which is driven by meaningful practical human needs, the Reconfigurable House is simply an ambient audio-visual event, with open-ended behaviors and subject to the improvisational creativity of its visitors. On the one hand, it is an interesting project because the functionality of its low-tech components has been reoriented and hacked to perform alternative functions; on the other hand, its value lies in that it demonstrates the idea of managing tangible computing environments by physical – not only digital – synapses.

In this hypothetical perspective, the physical components of ambient environments could be combined by creating physical synapses with each other (either through physical connection, contact or topological proximity), which would mean different events and functionalities, predetermined or unanticipated. The physical form of these components could be shaped in such a way as to provide sufficient information to users about their possible synapses and the way they can be connected. The Tangible Media Group at the MIT

Media Lab[30] has been using these ideas in their research projects since the 1990s. The researchers of the group have been exploring ways for people to interact with digital information directly though human physical skills – not through graphical user interfaces, Windows OS, a mouse and a keyboard. In this case, digital information acquires physical form, and is embedded in real objects that represent and control the information, providing the potential for tangible interactions with digital space.[31] This line of research follows Mark Weiser's vision of invisible ubiquitous computing, but extends it to establish a new type of interface between the physical and the digital. This is what the group calls a Tangible User Interface (TUI), namely everyday physical objects digitally augmented with graspable qualities and rich intuitive affordances.[32] In this case, meaning is conveyed through the interaction of users with these objects, which is not different from what happens in everyday interaction in the physical and social world.

Open Potential Environments

The functionality of end-user driven intelligent environments, like e-Gadgets, is the result of the interaction between people and their heterogeneous components that cooperate to optimize the behavior of the system. Therefore, the actions of users in such environments are not always successful in terms of their functional outcome, because this outcome depends on the dialectical relationship, the "conversation," between users and the technological artifacts that constitute the environment.

Users of e-Gadgets can actively customize their environments through the provided technologies, which are functionally flexible and physically tangible. They can creatively configure the system with unexpected synapses, connecting compatible or incompatible e-gadgets, in order to perform functions that have not been predetermined: "This approach regards the *everyday* environment as being populated with tens even hundreds of artifacts, which people (who are always in control) associate in ad-hoc and dynamic ways."[33] Users who know the constraints and affordances of these artifacts can freely create ad-hoc synapses and see whether they like them or not, whether they are as functional and useful as they prefer. This cannot be anticipated, because only by trial and error will they be able to judge which of them are useful and efficient enough for their living habits.

Therefore, according to Kameas, the creation of certain synapses does not necessarily imply successful functionality, in the same way that people do not always achieve their goals in everyday cooperative human interaction. He thinks that we have to get used to the idea that "smart" systems may sometimes fail to meet our wishes, and this is exactly because they are smart and because the system is not pre-programmed to anticipate all possible scenarios and respond effectively to all possible circumstances. By leaving the potential failure of the system open, its capacities for adaptation are extended. Just like the outcome of human conversation, which cannot be anticipated, the outcome of interaction between users and environment is indeterminate. This asymmetrical interaction echoes activity theory: users, as creative subjects, create their personal ad-hoc synapses, driven by their indeterminate intentions and desires.

Communication in the e-Gadgets system between users and artifacts, and between artifacts, is assisted by the middleware, which facilitates the negotiation of functionalities between devices:

> The solution we propose is communication (hence the term "extrovert"), as opposed to mere message exchange… In the e-Gadgets approach, a Synapse is formed as a result of negotiation among eGadgets. Negotiations and subsequent data exchange are based on ontologies possessed by the eGadgets; the only intrinsic feature of an eGadget is the ability to engage in structured interaction. Such environments are "ecologies" of interacting artifacts, processes, human users, infrastructure and software, demonstrating open-ended, indeterminate and multiple functionalities.[34]

The environment is here conceived as a symbiotic "ecology"[35] of autonomous heterogeneous agents, where artifacts and people collaborate and interact, producing unpredictable emergent behaviors. Like the entities of a complex biological system (such as a swarm), the components of the ecology have no internal representations of the other components or the environment in which they are embedded. As individual entities, they preserve their autonomy and their goals, and they demonstrate emergent behavior and intelligence.[36] This, according to Zaharakis and Kameas, provides flexibility, coherence and adaptability to the system: "[a]lthough the entities will not have explicitly represented models of the world or of the others the emergent ecologies will unfold coherent collective behaviour based only on the entities' own agenda of actions and their intrinsic inclination to preserve their own goals."[37]

As mentioned in Chapter 3, swarm systems are "closed" and interact on the basis of predetermined simple rules. To change these rules in a technological swarm simulation, the programmer must intervene. In the e-Gadgets ecology, however, the rules can be changed by users themselves. Moreover, as suggested in Chapter 2, swarms are homogeneous systems of either biological or technological units, not social. But e-gadgets are characterized by heterogeneity, because they comprise both artificial and human agents (such as users and designers).

The agents are autonomous "construction units," with a set of "primitive" behaviors that interact. They comprise sensors, actuators, microprocessors or their combinations, modeling a neural network capable of learning and evolution. Sensors and actuators here represent the input/output neurons, while the computational units and the synapses represent the hidden neurons and the local interactions of organisms in the supposed "ecology," respectively. The group of artifacts can be considered to be a network of sensors with a limited power, computational capacities and memory. Based on their limited capacities for processing, instead of sending simple data, they can make local simple computations, and transmit only the required and partly processed data. They have a physical and a functional identity, which they can publicize in order to communicate (with other units or the user). The "ecology" is constituted by the relations between the components, the functionality of these relations and the emergent and indeterminate overall behavior, which is the result of a common consensus (a common view of the environment). Thus, by combining the sensor network with computationally augmented artifacts we can trigger constant formation of new "societies," which can provide emergent behaviors that did not exist in the individual units. In all cases, the computational capacities and the intelligence of the "ecology" are distributed in the central nervous system, the peripheral system, the materials, and the natural phenomena created by the interaction of the ecology with the environment.[38]

Zaharakis and Kameas propose an extension of the user-driven functional level of the behavior of the e-Gadgets system, as described above, to include a proactive "social" level. This "social" level involves interactions between the components that disregard the user; this makes the system able to adapt optimally to unpredictable situations.[39] It produces behaviors of social intelligence, which are programmed to be hierarchically prioritized in relation to the behaviors of the functional level, thus allowing the e-gadgets to perform the most socially rational operation, according to the given situation. The system uses a selection mechanism to choose between the two levels and to secure independence in determining the local situation and response, while achieving a socially driven behavior of the e-gadgets. Some examples of basic functional behaviors include: "turn light on," "produce a specific sound," "move towards a specific direction" and so on. These behaviors depend on the actuators and the publicized synapses of the e-gadgets that determine the affordances in the physical and digital space. At the same time, the system supports basic behaviors, such as "form synapse," or "learn," to ensure the composability and changeability of the system.[40]

Users can compose an e-GadgetWorld, say morning awakening, through plugs and synapses according to their personalized preferences for their morning awakening and the available e-gadgets: this world will include an alarm clock, bed-mattress, a bedside picture frame, a bathroom mirror, window blinds and a room light. They can program it so that the light intensity will gradually increase after the alarm ring goes on, while soft music starts to play and images of summer holidays are displayed on the bathroom mirror, before broadcasting the daily news.[41] Yet, to optimize adaptation, the system will use not only the functional, but also the "social" level, in order to adjust its functionality to failures of the synapses. The "social" level supports behaviors inspired by social studies, such as "benevolence," "non-intrusion," "altruism," "responsibility," "antagonism," "empathy, "emergency" and so on. These behaviors are structured according to Brooks's subsumption architecture (see Chapter 2), which organizes levels of action according to hierarchical priorities: higher level social behaviors subsume lower level behaviors. In this way the system acquires sufficient autonomy to be able to optimize problems and malfunctions that cannot be controlled and predetermined by the users.

In the morning awakening example, the social motive "selfishness" means that all e-gadgets will keep prioritizing the achievement of their functional aims (such as lifting the blinds). The social behavior "social responsibility" means that, in case the devices fail (if, for instance, they cannot lift the blinds), the e-gadgets will try to find another way to perform the function (for instance to rotate the blinds, or to increase the artificial light intensity in order to compensate for the lack of natural light). Other similar social behaviors that optimize the functionality of the environment are: "altruism" (when the system attributes services to other e-gadgets and cooperates for the achievement of better results), "non-disturbance" (gradual and thorough change of the space, so that the users are not disturbed), "emergency," etc.[42] The resulting intelligence emerges not only from the interaction between user and system, but also the interactions within the system. Thus, intelligence is not located within isolated agents, but is perceived within the shared space of agents comprising the system, an idea that goes back to Brooks's definition (see Chapter 2) and Glanville's cybernetics-inspired argument about intelligence (see Chapter 4).[43]

Environments like the e-Gadgets provide the means for creative use within a wide range of choices. Depending on the variety of devices that can share their services, and the number of embedded sensors and actuators on furniture and architectural elements, the system can amplify the creative indeterminate possibility for change. These considerations point to a conceptual construction, what we will call "open-potential environment." Designers and engineers can implement this environment, which is adaptable to the needs of its users, by means of tangible user-driven systems. But apart from the creative and improvisational actions of users, who can adjust its operational rules, this environment employs levels of functional autonomy. It is important to acknowledge the fact that augmented architectural space acquires properties beyond the traditional capacities and functions of architecture. The principles and technical possibilities of user-driven ambient intelligence and tangible computing applications, point to an alternative conception of architecture; we can regard architecture as a "virtual machine" (see relevant discussion in Chapter 1), or in other words, a *machinic assemblage* of underdetermined open-ended functions and emergent behaviors.[44] Physical space, in this sense, is a framework with embedded digitally augmented physical objects that relate to this framework in terms of their location and function. Thus, architecture can be defined as an "ecology" of interacting technosocial entities – space, artifacts, human users and software – demonstrating open-ended, indeterminate, multiple and user-customized functionalities.

This assemblage is potentially open to the needs and preferences of its users, rather than system designers, who can cooperatively improvise to determine the rules of its operation. Sometimes these rules and operations can be surprisingly unexpected like, for instance, when users connect the closing of a door with the pausing of incoming phone messages, or the opening of the window blinds with the execution of a series of musical scores in a certain priority sequence. More interesting actions could be proposed; for instance when the bell rings or the camera records a visitor, the TV signal is modified. Thus, by means of end-user driven "open potential" environments, users can imagine and implement endless possible domestic functions, which may or may not make sense.

Living in Open Potential Environments

Spaces that combine physical and computational properties prompt us to consider a more direct and embodied interaction, engaging real-world social behaviors and new forms of interactional meaning. This implies alternative perspectives, and a move from spaces that are responsive to practical and everyday needs, to tangible configurations and environments designed for social interaction.

Alternative Perspectives

Although "tangible computing" and "open potential" environments seem to suggest interesting prospects for flexible and adaptive environments, the functional capacities of user-driven or autonomous intelligent environments are rather conventional, because they are limited to domestic assistive services. Maybe this is because studies in intelligent environments have mostly addressed domestic space, as the majority of research in ambient intelligence shows, with an emphasis on anticipation of everyday needs and personalized automation. The question that arises is whether these systems could be enhanced with alternative capacities, beyond the provision of simple or complex conventional domestic services.

If we look back at the work of post-war visionary architects like Archigram, Constant and Price, we will see a different approach from that of contemporary intelligent environments, in terms of privacy, control and communication. Archigram envisaged free and flexible environments operating by means of cybernetic technology. Constant had a much more radical idea in his New Babylon project, which expressed a rejection of the very concept of privacy and personal or family dwelling. His work addressed playful atmospheres, collectivity and temporality. What pervaded much of this work was the need for human communication and social interaction – beyond the fulfilment of merely practical necessities – which would be implemented by the shared use of mobile architectonic elements. This social dimension of Constant's New Babylon meant that architecture was not in itself a social agent, an actor, or a living entity, but a mediator of social relations.[45] Like Constant, who saw the environment not as a functional space but rather as a catalyst for social interaction, Cedric Price emphasized the social and communicative potential of his Fun Palace project, which, by means of tele-technologies and networks, would enable human participation in open discussions and democratic conversation with educational purposes.⁴⁶

Present-day recurrence of such digitally-mediated alternative forms of social interaction have not only populated internet and mobile applications;⁴⁷ several researchers have also considered this possibility in the context of intelligent environments and tangible computing. For instance, the Tangible Media Group at MIT has explored these ideas through the augmentation of tangible things with interactive properties that enhance abstract and playful communication between users, prompting them to develop their own ad hoc communication mechanisms. The project *InTouch* is an interactive piece that examines the possibility for non-verbal communication, mediated by a tactile mechanism of three rotating cylinders, embedded with sensors and actuators, in two interconnected units; when a person rotates the cylinders of one of the units, information is transmitted to the other unit in the form of tactile sensations on the other person's hand.⁴⁸ Technology here participates in interpersonal communication, without the need for a symbolic language. Meaning is not encoded in the system, but is rather produced in embodied tangible interactions, in use, and in the users' coupled relation with that technology.⁴⁹

Tobi Schneidler has also explored this possibility of non-verbal interaction in his *RemoteHome* project, an installation that addressed tactile communication between two people sharing a flat that existed at the same time in two distant cities – London and Berlin. Tactile stimuli were provided by computationally actuated physical pieces of furniture (such as a moving bench and a vibrating wall).⁵⁰ Along the same path, researchers in the context of the *Ambient Agoras* research program have implemented projects to turn everyday spaces into public "agoras" for the exchange of information and ideas. They designed projects such as the *Hello.Wall®* and the *ViewPort®*, which can display ambient information in intermediary public spaces in workplaces (such as a public corridor, the foyer, or the cafeteria), in order to mediate human social interaction.⁵¹

All these projects suggest the use of tangible computing systems to mediate social contact in everyday environments, beyond the mere fulfilment of practical necessities, domestic services and needs, sometimes going beyond the acceptable limits of private and public life. The architecture firm Hariri and Hariri, for instance, suggested the possibility of using plasma and liquid crystal walls in the speculative *Digital House* as ambient displays, to communicate messages to the neighborhood, thus allowing public life to potentially interfere with the private life of the house.⁵² For Schneidler, the most important thing in the incorporation of ambient intelligence in physical space is the possibility to create new cultural forms of inhabitation, as well as novel methods for social relations. Referring to the RemoteHome he adds that "It's very important for me that the project is not about interactive technology or smart space per se, but about creating environments that act as mediating devices for a social statement."⁵³ Architectural spaces are thus "quasi-objects," what Papalexopoulos calls a multiplicity of changeable, time-based "virtual" conditions that mediate the relations between users.[54] In this case, the digital entities that are embedded in architectural space can participate in its communicative capacity, provided by pervasive information technology.

Intelligent services and pervasive technologies are involved in our daily routines, actively adjusting their operations according to our activities. By affecting matters of domestic security, privacy and comfort, and extending or controlling the boundaries of our spaces, they may cause certain changes in the way we understand domesticity and experience the home. It is therefore important to acknowledge a potential shift in our habitual dwelling practices, as living and working spatial and temporal patterns are redefined. In the remainder of the chapter we will look at these important issues in AmI; the complex interactions between space, infrastructure and the people that use and inhabit it, pertaining to temporal and spatial questions about the settings in which it is deployed, and the meanings that are produced, particularly in relation to space and place.

Space, Place and Inhabitation in Ambient Intelligence

The "open potential" environment deploys energy and information services in domestic space; it does not change architectural space per se, nor its physical boundaries and layout. However, space can play a role in determining the configuration of the system. The physical features and constraints of space, its type and geometry, what Malcolm McCullough calls *persistent* spatial structures, can contribute to the optimization of pervasive systems, despite the emphasis on dematerialization that, as discussed in Chapter 1, characterized certain stages in the experimental work of post-war architects.[55] Spatial configurations can play a role in coordinating communication between devices, and the spatial modeling of the active digital activities.[56] As Malcolm McCullough explains, "In contrast to the usual assumptions about formless dematerialization, the rise of pervasive computing restores an emphasis on geometry."[57] In other words, ambient intelligence entities are not indifferent to the building type and its physical configuration, because they are located *somewhere*, in a specific place. The potential activities and services provided by digital systems are based on a range of recognizable situations, and on the activities and functions of the space in which they are embedded.[58] The type and functions of space, its furniture and its devices can considerably influence the design and distribution of ambient intelligence entities. Designers should consider their distribution in relation to the existing and possible functions of space. McCullough has proposed 30 such situations – places of activity rather than spaces – for work, inhabitation, entertainment, and transportation: places for presentation, cooperation, learning, observation, sheltering, relaxation, service provision, meetings, socialization, shopping, sports, temporary residence, driving, walking and so on.[59] These situations correspond to different types and configurations – spatial distribution and constitution – of pervasive technologies.

This view brings forth our constant need for permanence – the fact that our spatial structures and buildings should provide the enduring periphery within which people can move and customize their environments. Discussing the role of architecture and the built environment in the digital world of constant mobility and flow, McCullough argues that architects must provide a fixed ground for the movement and exchange of information, namely ambient intelligence.[60] He explains that architecture should be "calm" and address "the affordances of everyday life rather than fashionable statements in form."[61] Architects should regard a building as the physical "ground" within which digital systems and operations can be "absorbed."[62]

Frank Duffy, ex-president of the Royal Institute of British Architects, analyzed buildings into four layers according to their degree of permanence: the structure, which he called Shell and is the most permanent layer, the Services (the plumbing, cabling and air-conditioning systems), the layout of partitions, which he called Scenery, and the most changeable layer, the furniture, which he called Set. To these layers of change, Stuart Brand added another two: the Site, the geographical and urban location of the building, and the Skin, the exterior surfaces which change every 20 years or so.[63] Similarly, Bernard Leupen reviewed a number of mainly 20th-century buildings, to analyse them according to a system of five building layers (structure, skin, scenery, services and access), with different rates of change. But by contrast to Duffy, Leupen, who built on John Habraken's classic work on mass housing,[64] argued that any of these layers can acquire the role of the permanent layer, what he called "frame," either alone or in combination with others.[65] Following such an analysis we can suggest that the digital system can be embedded within this hierarchy of permanent and less permanent layers of organization, like so many other technological layers in buildings were absorbed in the past and throughout the history of architecture. Rem Koolhaas recently discussed the historical aspect of this embedded layer of intelligence in buildings. He became practically aware of its extensive presence in architectural components while working on the 2014 Venice Architecture Biennale. Examining the history of architecture at the scale of its individual elements (walls, floors etc.) through microscopes, he and his team became increasingly aware of the constantly accelerating penetration of "smart" devices, sensors and systems in architecture.[66] Thus we can suggest that ambient intelligence and its units constitute an extra "digital" layer, an assemblage of "intelligent" devices and software – agents with input, output and communication plugs. These agents can respond to the changeable needs and desires of occupants, through the management of information and energy supply within the domestic environment. In this case, the "digital layer" would be the changeable part within the framework of Leupen's analysis, while the physical spaces in which it is embedded would constitute the "frame," the permanent part (Figure 5.5).

Using the techniques of end-user driven intelligent environments, the components of this "digital" layer would be functionally "open" and adaptive

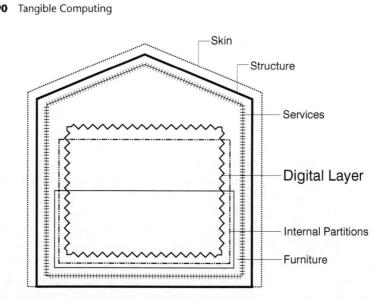

FIGURE 5.5 The "digital layer" in the system of layers analysis of buildings

mechanisms, able to be programmed and reprogrammed by users, facilitating customization and user-driven system behaviors. This would involve changes in the personal sense of comfort, the ambience and the sensory experience, achieved by information and energy management devices that are adjusted by the occupants of the domestic environment.

Dimitris Papalexopoulos, on the other hand, proposed a more dynamic relation between ambient intelligence and architecture. Inspired by the biological metaphor of swarms, and its use in projects such as the e-Gadgets, Papalexopoulos argues that the distribution and the cooperation of ambient intelligence entities in space is a new paradigm in architectural design, because it suggests that functionality is changeable; spaces can acquire multiple functions, determined and pre-programmed only on the basis of the digital entities and the actions of human agents. But unlike Oosterhuis, who adopted the swarm concept to propose the constitution of physically transformable structures (in which the interaction between units produces overall transformations),[67] Papalexopoulos suggests that the swarm logic can be incorporated into existing built space, which does not mean a new *kind* of architecture.[68] Like McCullough, he thinks that designers of pervasive digital systems in intelligent environments should redefine the relation between architecture and digital technologies: "the functionality of an ambient intelligence system depends on the design of the space that is supposed to host it."[69] Papalexopoulos coins the term *digital territory*, a temporary space of ambient intelligence, which defines its relation to physical space by means of *bridges* connecting the physical and the digital (such as a sensors, actuators and RFIDs). The inhabitants of ambient intelligence environments can change the structure of these bridges, which are open to transformations according to human needs. Designers of ambient intelligence spaces should then try to find methods to connect the digital units and their distribution, as well as define the conditions pertaining to the extent of communication between the units and the environment. This means that part of the design process would entail the creation of "location diagrams" that determine the distribution and position of the bridges connecting physical and digital entities. Design processes should also determine what information is gathered from the environment and processed by those bridges, the kind of network that connects them, and who defines these parameters.[70] Although the sphere of influence of these units and their interconnections, the digital territory, does not coincide with the place in which they are embedded, place *does* relate to them.[71]

For Michele Rapoport, however, ambient technologies, and in particular domestic surveillance systems in smart homes, play a critical role in how they alter notions of place and space. In intelligent domestic environments, the idealized notion of the home as a private respite and a "corner of the world," as Gaston Bachelard would have it, is challenged by the fact that its boundaries are penetrated by the very technologies that are supposed to protect them. The concept of *place*, which is contingent upon particular locales and situated subjects, and implies secure barriers, defendable enclosures and bounded configurations, is opposed to *space*, which implies the general and the everywhere, the geometrical and the measurable, and suggests freedom and safety through surveillance. This Heideggerean conception of space as a quantifiable, measurable entity, lacking an essential relationship with specific locales, subjects or other things, is what distinguishes it from place, which also resonates with the view taken by geographers like David Harvey and Doreen Massey, who have emphasized the relational aspect of space.[72] Intelligent environments, Rapoport continues, facilitate the notion of space rather than place; the idea of the home as *place* is destabilized as it diffuses into monitored images and deterritorialized timeless spaces, seen from practically everywhere, and detached from concrete sites and situations, bounded by incorporeal informational barriers. Tangible places and locales turn into abstract spaces and virtual "data doubles," while bodies are transformed into cyberspace simulations and timeless subjects.[73]

In Dourish's discussion of the distinction between space and place (which is based on several references from architectural theory), on the other hand, the concept of place is central to the design of computational environments; within computational environments, place pertains to social activities. As he explains, the distinction between *space* and *place* in the context of computationally augmented environments, highlights the distinction between the physical and the social. While *space* refers to physical configurations and settings encompassing properties like proximity, orientation, position, direction and partitioning, *place* refers to the way these settings are invested with meaning and understandings of behavioral appropriateness, through the way social relations and events unfold; thus place is the behavioral framing of an environment according to specific social norms.[74] And since place is linked to the activities that may happen in computational environments, space, as a physical structure (or even as a "real world" metaphor in virtual environments), is not necessary for the establishment of meaningful social interactions, unless it contributes directly to those activities.[75]

Steve Harrison, in a paper co-authored with Dourish in 1996, argued that space is hierarchically prior to place; place is a reality appropriated and enacted by users that arises and develops from practices within pre-given spaces.[76] Ten years later, however, Dourish reviewed and reflected on this early paper to reject the implied duality between space (as a stable pre-condition) and place (as dynamic and behavioral). This early view was not aligned to the idea that space might also be a dynamic aspect of spatiality, which draws on Michel de Certeau's more elaborate distinction between space and place. De Certeau wrote that "... space is a practiced place," thus associating place with stability, and space with social practices, trajectories and vectors.[77] Dourish went on to argue that both space and place are products of embodied social practice, mutually constitutive, albeit different. Places, which are experienced first, involve the cultural and social foundations in which we encounter heterogeneous locales with local meanings and singular properties. Spaces are the outcomes of such encounters, and this is particularly evident in the way wireless and ubiquitous technologies are already spatializing practices. But that is not to say that technologies create new forms of electronic space (as, for instance, in the use of mobile phones in the city). They cause and open up new forms of social and cultural practices, making users re-encounter sites of everyday action, experientially transforming existing spaces and the ways in which the extents and boundaries of space become understandable, practical and navigable.[78]

This understanding of space within the technological environment is particularly evident and relevant to several technologically meditated practices, performed in the context of the city by means of location-aware mobile communication systems, the so-called *locative media*.[79] Because of the emphasis on urban drifting and the redefinition of the dominant spaces of the city, in many instances, researchers have considered these practices to be echoes of the 1950s situationist urban strategies, such as the "dérive" and "psychogeography,"[80] augmented with (and enhanced by) an additional digital layer.[81] Locative media projects such as *Tactical Sound Garden toolkit*, *You are Not Here*, and *Walking Tools*,[82] seem to re-enact ways of urban habitation; they produce a new kind of spatiality through technologically mediated, embodied and situated playful actions, sensory perception, memory and emotional reaction to specific locales, as opposed to the homogeneous, metric space depicted in geometric representations and maps.[83]

Similarly, domestic ubiquitous technologies may not change the structure, form or geometry of physical space, but they do influence domestic life in terms of its social and moral organization; patterns of communication, rhythms of daily routines, the boundary between inside and outside, leisure and work time, the way these activities are separated, and the rhythms that control the movements of people and objects over these boundaries.[84] Thus, the domestic, as a form of cultural practice in the context of ambient intelligence, should not be assumed as a pre-given and stable category, but rather as actively produced and enacted within everyday practice.[85] As Paul Dourish and Genevieve Bell have argued, we have to consider conceptual devices and standpoints capable of reframing technological discourse about the home. Such an approach would question the way technologies are domesticated and by whom, the new gender relations they produce, and how boundaries and separations are maintained. It would also suggest the possibility to introduce "tinkering" with technologies (a practice usually associated with dirty and marginal rooms that are not part of the main house), and end-user programming – a sort of messing around with technology.[86]

Such practices are relevant to what science and technology studies researchers, as well as media and communication theorists, have termed "domestication." A metaphor for "taming" an animal, domestication is the "housetraining" process of integrating "wild" technologies into the everyday structures, daily routines and values of users, particularly those in the domestic space.[87] It deals with the cultural, social and technological networks of everyday life. In particular, domestication refers to processes of utilitarian and emotional appropriation of, and adaptation to or rejection of, technological artifacts in the household by their users. The household in these studies is considered as both a social and economic entity, constituting a sort of moral economy in constant exchange with the economy that surrounds it in the public sphere.[88]

The domestication is thus a movement of artifacts from production to consumption, segmented into a number of moments from the formal market economy of the public sphere to the integration of technologies in the patterns, habits and values of the private sphere of the household. The first step is production and advertisement of artifacts, which involves imagination and *appropriation*, and the bringing of technologies into the domestic environment. The next step, called *objectification*, is shaping technologies to merge with the physicality of the household. These steps are followed by *incorporation*, the remaking of the meanings, values and norms of the artifact, and finally *conversion*, their transfer back to the outside world.[89] The third of these steps, *incorporation*, points to the way technologies are used by consumers/users, which may be different from the designers' and marketers' intentions. In other words it addresses the negotiated space between the designers' views and users' needs or interests, and between what is *inscribed* in the object and how it is *de-inscribed* by users. At the same time, domestication is characterized by mutual change and adaptation, because users not only modify their artifacts according to their needs and desires, but they are also transformed by them, in behavior and emotion. Most importantly domestication pre-supposes "that users play an active and decisive role in the construction of patterns of use and meanings in relation to technologies."[90] These concerns reflect what the *open potential environment*, as depicted in this chapter, is all about, because it deals with technologies that are tangible, enacted and customizable, namely, open to the wishes and desires of its users.

At the same time, pervasive technologies can induce changes in the spatial and temporal organization of the household. An example can be considered in Katie Ward's study of the "home-work" use and "domestication" of the personal computer in the household, based on ethnographic accounts in a coastal town in North County Dublin. In Ward's thinking, the home is both a material and a symbolic entity, constituted by the web of human relations and the values and habits of the family that the home, in turn, articulates, which is the basis for the creation of a household identity.[91] As "alien" technologies come into this private sphere of domesticity, users inscribe them with private meanings and symbolic value. They redefine and reshape them to conform to established routines, patterns and social hierarchy. Thus, artifacts are consumed and appropriated – domesticated – in a two-fold process of negotiation, conflict and collaboration with artifacts. The domestication of the computer and internet media, in the study of Katie Ward, involves not only its appropriation by users but also the production and reformation of domestic arrangements, temporal routines, as well as habits and behaviors in the household, in the same way that the introduction of the TV in the household induced the employment of supporting technologies and the creations of new spaces (such as TV lounge, TV dinner etc.).[92] Ward focuses on the objectification and incorporation stages of the domestication process, which was not as straightforward as it seemed, to demonstrate the untidiness and the blurring of the boundaries of these stages, as technologies, in her study, were not only subject to reconfiguration by their users, but would also disrupt temporal routines and spatial arrangements. As work at home sometimes confuses the boundaries between family matters, leisure time and work life (hours spent in the home and hours spent doing work), users felt the need to manage the intrusion of these technologies by clearly demarcating the boundaries between home and work in the household, in both temporal and spatial terms. For instance, they would locate their computer on a different floor or distinct room to separate it from family life, assign different meanings and use patterns to the computer used for work and the computer or the TV used for entertainment, or install separate phone lines to separate home- and work-related communication.[93]

Therefore, pervasive technologies in the home not only depend on the physical configurations in which they are embedded and in the context of which they are used, but also affect the state of affairs of the household in terms of spatial arrangements, domestic life patterns and everyday rituals. There seems to be a potential ground for research into these changes that will need more relevant case studies and ethnographic accounts.

Overview and Conclusions

Throughout this book we have made an effort to explore several questions pertaining to the computational augmentation of built space, with particular emphasis on domestic environments. Our investigation was driven by the implied capacity of digitally driven structures, interactive installations and intelligent environments to adapt to changing human needs and desires. We looked into the historical background and addressed relevant insights from the social studies of technology, the history and theory of artificial intelligence, the theory of human–computer interaction and post-cognitivist theories; we also discussed functional and technical issues, as well as psychological and cultural dimensions. We thus assessed the past and present state of the art, what the future might be and what relevant future research may involve for engineers and architects. At the same time, we highlighted concerns about the effects on the patterns of domestic life, as well as spatial and functional transformations.

Starting our investigation from the recent past, we have shown that post-war architects addressed cybernetic concepts, such as information feedback, self-regulation and adaptation. We conclude that Archigram's and Banham's iconographic and theoretical approach respectively, and Price's, Friedman's and Zenetos's more elaborate and thoughtful designs, suggest a shift from flexible, functionally determinate structures to adaptive and user-determined environments assisted by information and communication technologies. These proposals, which were part of a wider cultural and epistemological shift in thinking that led to our contemporary digital culture, were harbingers of a later interest in computationally augmented architectural spaces. Since the early 1990s, a number of architects and computer engineers have produced publications and experimental projects that generally suggest a future in which spaces, particularly domestic ones, could embody intelligent systems, mechanisms and services, in order to adapt flexibly to the demands of their users.

This discussion led to the second chapter in which we examined the historical background of artificial intelligence – from the classical and the connectionist approach to the new AI (mostly applied in robotics and artificial life applications), multi-agent systems and swarm intelligence. But its most prevalent application in the built environment was the experimental project that emerged in the field of the so-called Ambient Intelligence. This is a vision in computer science that promises a future of automated services in houses, able to learn by experience, and proactively enhance and anticipate our everyday activities by modeling users, their habits and daily activities, to construct rules to determine their functionality. Intelligent environments, like the iDorm and the PlaceLab, are able to automate domestic functions and services (which sometimes may cause feelings of insecurity and loss of privacy), and adapt dynamically to changing circumstances that their computational system has already predefined. As opposed to these kinds of environments, which operate by means of autonomous agents that collaborate to anticipate human needs, another strand of intelligent systems in houses, the so-called end-user driven intelligent environments, promise a future for more open-ended adaptable spaces.

Such environments, along with other computationally augmented structures, were then discussed from a psychological perspective, leaving aside their functional capacities. Since they are endowed with living characteristics – and behaviors like perception, action and intelligence – they can also be evaluated in terms of their tendency to question the boundaries between the animate and the inanimate. In the third chapter we examined these psychological dimensions from the perspective of experimental psychology, anthropomorphism and animacy perception in psychophysics, as well as the social studies of science and technology. We deduced that certain features of form and motion in mechanically changeable and intelligent structures (such as Hyperbody's Muscle projects, Jansen's Straandbeest, and Philip Beesley's Hylozoic Ground) can perceptually blur the boundaries between the animate and the inanimate, an idea that points to Turkle's concept of "marginal" objects. Considered in the light of the history and theory of artificial life (which includes living machines from 18th-century automata to contemporary sociable robots), and the suggested constant historical exchanges and associations between machines and organisms, these architectural productions are not only another expression of the long-standing biological metaphor in architectural discourse, but also concretizations of a wider cultural (and perhaps psychological) drive to overcome the boundaries between human and machine, subject and object, or human user and architecture.

This is aligned to the idea of interactivity as a cooperative process involving user participation, which was examined in Chapter 4. Instead of defining interactive architecture using insights from architectural examples and publications only, we looked deeper into the emergence of the concept of interactivity in the post-war cybernetic discourse, which led to contemporary notions of interactivity in the context of the new digital media and human–computer interaction. Furthermore, drawing on conversational, situated and embodied models of interaction, social and post-cognitivist theories of technology (actor–network theory, distributed cognition and activity theory), and theories of technology that use insights from the phenomenological tradition (the post-phenomenological approach to technology), we attempted to define the theoretical ground in which interactive environments can be contextualized and evaluated. Thus, although installations like ADA and Performative Ecologies are considered to be interactive, they hardly allow users to intervene and customize their operations. Interactivity in adaptable domestic environments should favor participation and open-ended "conversation," rather than autonomous functioning.

Therefore, in the last chapter, we examined an alternative path to human–computer interaction in domestic environments, using ideas from tangible computing and end-user driven intelligent environments. E-Gadgets and the Reconfigurable House allow users to creatively customize the operational rules of the system and synthesize behaviors. This led us to propose a conceptual construction that we termed an "open potential" environment, which should be able to address functional indeterminacy and open-endedness. Of course, these capacities are constrained by the very devices that participate in the system (both their physical and digital functions), and the fact that change is produced by the processes and functions of the intelligent system and its connected devices and appliances. Therefore, although architectural space does not change physically but only in terms of its embedded functions, it can still be considered to be a flexible environment, because it acquires a multi-functional capacity through the open-endedness of the embedded system.

Technically, the capacities of the "open potential" environment can be achieved by means of friendly interfaces and the necessary middleware that will be able to manage and represent, in common terms, the functional capacities of heterogeneous devices, providing exchange of information between incompatible devices, sharing of their identities, and interpretation of messages accordingly. Thus, users can understand these identities, and interconnect devices to execute novel and indeterminate tasks. The result may pertain to changes in the experience and functions of the physical space: changes in lighting, openings, air-conditioning, the performance of audiovisual devices, changes in informational systems, and so on. In case users require objects to be able to move, the environment may embody robotic applications, such as those in the PEIS Ecology project, in order to serve its inhabitants.

In the "open potential" environment, users retain their agency as intentional subjects and can creatively engage with the functions of the ambient environment. As discussed in Chapter 4, this is grounded in activity theory and the theories of embodied interaction, which favor intentional actions of situated embodied subjects. Subjects are materially engaged with the environment, through tangible artifacts. As discussed previously, user involvement in determining the functions of the system, is not unidirectional. The devices of the system can also make decisions and take action to co-shape the environment, by activating what Zaharakis and Kameas called the "social" level of the system. The system uses subsumption architecture to achieve its goal and override malfunctions and failures. It can make its own decisions according to a set of functional priorities, driven by social motives. Along with the users' improvisational actions, open potential environments can also act, on the basis of learnt experience, by means of cooperating agents, negotiating the success of their goal. The result is a symbiotic indeterminate "ecology" or rather a socio-technical assemblage, as Rapoport has argued,[94] engaging cooperative processes and non-symmetrical interactions (conversations), exchanges and emergent transformations between all involved agents (users and technologies, actors and devices).

In the "open potential" environment, action and agency is not solely attributed to users, but also delegated to the devices that make possible "programs of action." But this is not symmetry; symmetry would mean that artifacts act through mediations and delegations, and the gap between objects and subjects is overridden. In the "open potential" environment subjects and objects are not erased, or replaced by actants (either human or non-human), as an ANT perspective would suggest. Thus, Verbeek's post-phenomenological approach discussed in Chapter 4 is more relevant here, because it does not embrace Latour's concept of symmetry. The open potential environment in this framework points to the mutual constitution of humans and things, where artifacts, as mediators of human–world relations, actively co-shape the way humans act in and perceive the world.[95]

The implied materiality that pertains to the idea of tangible computing, as discussed previously, is also relevant to Verbeek's approach, which uses Heidegger's early phenomenological insights to account for artifacts that invite engagement with their material function, rather than their symbolic and iconic representations in the form of entities and events. Thus, unlike the e-Gadgets project, which was implemented by means of a detaching interface (editor tablet), the Reconfigurable House, as discussed, is much closer to the idea of material tangible interaction in ambient environments, although its actual implementation does not seem to provide meaningful experiences of inhabitation.

This relation between open potential environments (using tangible or other ambient intelligence systems) and domestic space, is another problem for further discussion and research. As we saw, ambient intelligence, both in public and domestic space, implies new forms of spatiality and questions traditional understandings of the digital as abstract, disembodied and placeless. Physical space is not indifferent to its embedded system, because it can be conceptualized as the stable "ground" that defines the way digital entities are distributed and function. Space, then, is seen through the concept of place, which pertains to concrete and situated activities, directly linked to the location and the functioning of ambient intelligence systems. The so-called locative media have demonstrated this interconnection between location and the digital. Also, socially oriented augmented applications and spaces, like inTouch and Remote Home, seem to manifest alternative notions of spatiality, as they assemble digital and physical spaces in concrete experiences. Therefore, thinking of ambient intelligence systems in relation to the physical aspects of domestic space can provide useful insights into the possible bridges and interconnections between the digital and the physical. In this sense, embodied and situated practices, as well as relevant "domestication" processes in augmented environments, can affect our daily habits and rituals, and hence our spatial and temporal configurations. Indeed, ethnographic research may provide useful insights into these correlations, and demonstrate that ambient intelligence, as an extra layer in domestic space, is critical for the design of our homes and the daily activities they shelter.

Notes

1 Dourish, *Where the Action is*, 16.
2 Ibid., 40–45. Pierre Wellner, for example, developed the Digital Desk at Xerox lab EuroPARC in Cambridge, England in 1991, to explore how office work could seamlessly combine both paper and electronic documents. The project augmented the physical affordances of conventional office desks (the physical world) with computational properties (using image processing software, projector and camera), making these two worlds work together and interact. See Peter Wellner, "Interacting with Paper on the DigitalDesk," *Communications of the ACM* 36, 7 (1993): 87–96. Also see Dourish, *Where the Action is*, 33–36.

3 Dourish, *Where the Action is*, 53.
4 Hardware components include batteries, a processor, sensors, and a wireless communication unit. Software implementation includes hardware drivers necessary to manage information exchange between the operating system and the external resources (sensors, actuators, screen etc.), a networking subsystem for the exchange of messages between artifacts, an operating system able to manage the use of resources and to translate service requests into commands to resources and vice versa, and also a middleware to provide abstractions that enable an artifact to be part of a distributed system.
5 Dourish, *Where the Action is*, 50–51.
6 Achilles Kameas, Irene Mavrommati, and Panos Markopoulos, "Computing in Tangible: Using Artifacts as Components of Ambient Intelligence Environments," in *Ambient Intelligence: The evolution of Technology, Communication and Cognition Towards the Future of Human–computer Interaction*, eds. Giuseppe Riva, Francesco Vatalaro, Fabrizio Davide, and Mariano Alcaniz (Amsterdam/Oxford/Fairfax: IOS Press, 2005), 122–123.
7 Traditionally, designers of interactive systems model the users and their potential activities in interaction, using techniques to evaluate the cognitive impact of different design choices on users, in order to apply those that better respond to the intended use of the artifacts. See Dourish, *Where the Action is*, 172.
8 Dourish, *Where the Action is*, 138–142.
9 Ibid., 172.
10 Ibid., 173.
11 Verbeek, *What things do*, 117–118, 136. Also Don Ihde, *Bodies in Technology* (Minneapolis: University of Minnesota Press, 2002), 106.
12 Verbeek, *What things do*, 217.
13 Ibid., 227–228.
14 Of course, as discussed in Chapter 2, the benefits of autonomous intelligent environments should not be overlooked, because there are "invisible" functions, like monitoring of infrastructure, networks and services, failure protection, security control and management of energy behavior, that humans alone cannot control. Intelligent environments that are capable of optimizing their energy performance and the conditions of space (both locally and in the overall building), as well as anticipate needs, and model and learn from users' activities to automate services and functions (which is important for people with special needs), are undoubtedly acceptable. In this case, however, the environment should appear reliable to users, and make them feel that they are in control. This need is also important for another reason: the possibility of anthropomorphism, namely the feeling that the environment becomes alive and takes autonomous action, which may cause feelings of anxiety and insecurity. This feeling may arise when, for instance, devices make unexpected activities.
15 Chin, Callaghan, and Clarke, "A Programming-by-Example Approach to Customizing Digital Homes."
16 Kameas interview.
17 Irene Mavrommati and Achilles Kameas, "Enabling End Users to compose Ubiquitous Computing applications – The e-Gadgets Experience," *DAISy Dynamic Ambient Intelligent Systems Unit*, accessed September 8, 2012, http://daisy.cti.gr/webzine/Issues/Issue%202/Articles/Enabling%20end%20users%20to%20compose%20Ubiquitous%20Computing%20applications%20-%20the%20e-Gadgets%20experience/index.html.
18 The GAS-OS facilitates the discovery of plugs of other e-gadgets, which publicize their properties by sending a message that contains all information that e-gadgets can have about each other. It also provides users with the capacity to activate and deactivate synapses, determines the compatibility of these synapses, and manages these synapses by sending notifications to other e-gadgets for changes in the situation

of other connected e-gadgets through the plugs. See Kameas, Mavrommati, and Markopoulos, "Computing in Tangible," 131–132.
19 These e-Gadgets were created by attaching simple pieces of hardware and software onto ordinary domestic objects (desk, chair etc.). For instance, the e-Desk was equipped with sensors to be able to sense weight, brightness, temperature and proximity. A pressure pad was placed under the top surface of the desk, whereas light and temperature sensors were equally distributed on the surface. Four infrared sensors were placed on the legs of the desk and another on its top surface. See Kameas, Mavrommati, and Markopoulos, "Computing in Tangible," 133.
20 Mavrommati and Kameas, "Enabling End Users to compose Ubiquitous Computing applications."
21 Kameas interview.
22 Kameas, Mavrommati, and Markopoulos, "Computing in Tangible," 135–136.
23 Kameas interview.
24 Dourish, *Where the Action is*, 16.
25 Ibid., 159.
26 Kameas interview.
27 "Reconfigurable House: Hacking Low Tech architecture," accessed October 12, 2013, http://house.propositions.org.uk. For a sequence of its operation see "Reconfigurable House," YouTube video, 4:04, posted by "Adam Somlai-Fischer," April 28, 2007, https://www.youtube.com/watch?v=ZZsEP08LWH4#t=67.
28 "Reconfigurable House," *Haque Design & Research*, accessed October 12, 2013, http://www.haque.co.uk/reconfigurablehouse.php.
29 "Reconfigurable House: Hacking Low Tech architecture."
30 See the Tangible Media Group website at http://tangible.media.mit.edu.
31 Hiroshi Ishii, "The Tangible User Interface and its Evolution," *Communications of the ACM* (Special Issue: Organic user interfaces) 51, 6 (2008), 32–36.
32 Hiroshi Ishii and Brygg Ullmer, "Tangible Bits: Towards Seamless Interfaces between People, Bits and Atoms," in *Proceedings of the ACM CHI 97 Human Factors in Computing Systems Conference*, ed. Steven Pemberton (Atlanta, GA: ACM Press, 1997), 235–236.
33 Kameas, Mavrommati, and Markopoulos, "Computing in Tangible," 140.
34 Ibid., 137.
35 According to Zaharakis and Kameas this consists of simple interacting units in synergy that, like a heterogeneous swarm, provide simple services and share their resources, thus maximizing the efficiency of communication, achieving complex and emergent overall behavior that is cohesive and functional: "As an engineering approach, SI offers an alternative way of designing intelligent systems, in which autonomy, emergence, and distributed functioning replace control, pre-programming, and centralization." See Ioannis Zaharakis and Achilles Kameas, "Emergent Phenomena in AmI Spaces," *The European Association of Software Science and Technology Newsletter* 12, (2006): 86.
36 As we saw in Chapter 4, this discussion can be informed by post-cognitivist theories and theories in the social study of technology (like actor–network theory, distributed cognition theory, activity theory and embodied interaction theory) that, despite their differences, seem to propose the conception of the word as a hybrid technosocial environment, consisting of heterogeneous associations, liaisons and synergies between humans and non-humans (tools/machines/artifacts/technologies). For an attempt to contextualize intelligent environments and adaptive architecture within these theories see Socrates Yiannoudes, "From Machines to Machinic Assemblages: a conceptual distinction between two kinds of Adaptive Computationally-Driven Architecture," in *International Conference Rethinking the Human in Technology-Driven Architecture*, Technical University of Crete, Chania, 30–31 August 2011, eds. Maria Voyatzaki and Constantin Spyridonidis (EAAE Transactions on Architectural Education No.55, 2012), 149–162.
37 Zaharakis and Kameas, "Emergent Phenomena in AmI Spaces," 93.
38 Ibid., 89–90.

39 Ioannis Zaharakis and Achilles Kameas, "Social Intelligence as the Means for achieving Emergent Interactive Behaviour in Ubiquitous Environments," in *Human–computer Interaction, Part II*, vol. LNCS 4551, ed. Julie Jacko (Berlin/Heidelberg: Springer-Verlag, 2007), 1019.
40 Ibid., 1024.
41 Ibid., 1027.
42 Ibid., 1027–1028.
43 Zaharakis and Kameas use an ecological metaphor here. Intelligence, in an ecology, is the result of a common culture, where beings produce social rules in order to evolve socially. Intelligence (by contrast to its traditional definition as an abstract concept pertaining to individual systems) can be examined only within the socio-cultural context of organisms. See Zaharakis and Kameas, "Emergent Phenomena in AmI Spaces," 85–86.
44 For a more in-depth discussion of these concepts in relation to adaptive architecture see Socrates Yiannoudes, "From Machines to Machinic Assemblages."
45 Sadler, *The Situationist City*, 149.
46 They understood that the epistemological shift from the structures and traditions of the industrial era to the post-industrial era of information technology in Britain, demanded an architecture capable of encouraging social transformation and proposing new patterns of knowledge acquisition that rejected the established systems of education and thinking. See Mathews, "Cedric Price," 91.
47 We could claim here that the Fun Palace anticipated current forms of virtual communities, where information is freely transmitted, in the context of the new interactive media and the internet. Although such a comparison may be simplified (because the Fun Palace has a physical architectural form whereas virtual communities are digital environments), the fact that the Fun Palace provoked this comparison is important. See Wilken, "Calculated Uncertainty."
48 See the inTouch webpage at http://tangible.media.mit.edu/project.php?recid=75.
49 Dourish, *Where the Action is*, 179.
50 The installation consisted of digitally enhanced actuated kinetic elements (a seat, a sofa-bed, and a wall) that facilitated the exchange of non-verbal and tangible information between the distantly residing "flat mates." The residents' sensor-recorded motions and activities were transmitted through the internet to the other resident's apartment, causing physical motions on the wall and furniture. For instance, when the London resident sat on the bench, a deployable sofa-bed in the Berlin apartment would be physically modified. In this way, as Schneidler thinks, the home was extended beyond its physical limits, and helped friends to stay literally in touch, by means of tangible and sensory communication – providing an emotional and intuitive form of their presence. See "RemoteHome," *Interactive Institute Smart Studio*, accessed May 22, 2014, http://smart. interactiveinstitute.se/smart/projects/remotehome/index_en.html.
51 See the *Ambient Agoras* website at http://www.ambient-agoras.smart-future.net.
52 "The Digital House," *Haririandhariri*, accessed December 2, 2014, http://www.haririandhariri.com.
53 Quoted in Lucy Bullivant, "Mediating Devices for a Social Statement: Tobi Schneidler Interactive Architect," *Architectural Design (4dspace: Interactive Architecture)* 75, 1 (2005): 76.
54 Papalexopoulos, *Ψηφιακός Τοπικισμός*, 102.
55 For instance, Archigram's proposal for the disappearance of architecture had ideological causes; they considered functionalism and the notion of type to be outdated, because all tendencies in society and technology of their time were concerned with flexibility and multi-functionality. They thought that the disappearance of architecture can liberate the environment from the "pornography" of buildings. See Cook, *Archigram*, 112–113.

56 Malcolm McCullough, *Digital Ground: Architecture, Pervasive Computing and Environmental Knowing* (Cambridge, MA/London: MIT Press, 2005), 98, 101–102.
57 Ibid., 101.
58 Ibid., 117–120.
59 Ibid., 118–141.
60 For McCullough stability assigns value to architecture. By contrast to the rapid motion of digital information through all sorts of information technologies, the quiet material permanence of buildings is a source of calmness. See Ibid., 63–64.
61 Ibid., 64.
62 Ibid., 172.
63 Brand, *How Buildings Learn*, 13
64 John Habraken, *Supports: An Alternative to Mass Housing*, trans. B. Valkenburg (London: Architectural Press, 1972).
65 See Leupen, *Frame and Generic Space*.
66 Koolhaas, "The Smart Landscape."
67 See Oosterhuis, "A New Kind of Building."
68 Papalexopoulos, *Ψηφιακός Τοπικισμός*, 156.
69 Ibid., 150.
70 Ibid., 154–155.
71 Ibid., 155.
72 Apart from philosophers like Heidegger and Merleau-Ponty, geographers like David Harvey and Doreen Massey have emphasized this relational aspect of space, which reflects a focus of their discipline on the spatial configurations that relate with, and affect, economic and social processes on a local and global scale. A relational view of space starts from the challenges posed on the concept of absolute space with fixed coordinates by the development of non-Euclidean geometries in the mid-19th century. This view rejected space as a fixed external container for human activities in a Kantian sense, and opened the way for the conception of space as a system of relations, a topological field influenced by activities and objects in space. See David Harvey, *Explanation in Geography* (London: Edward Arnold, 1969). Doreen Massey, writing in response to what she calls the effects of time-space compression on our notions of place and geographical uneven development, argued that the question of space pertains to how places are tied to both local and wider global relations and networks. A place, according to Massey, is not defined by physical enclosures and boundaries that separate inside from outside and an internalized history, but a unique point of intersection and conflict, comprising linkages to the outside and the global; it is a locus of networks of geographically differentiated social relations, consisting of processes, movements and communications. See Doreen Massey, "A global sense of place," *Marxism Today* 38, (1991): 24–29.
73 Rapoport, "The Home Under Surveillance," 325–326.
74 Dourish, *Where the Action is*, 89–90.
75 Ibid., 90, 149.
76 Steve Harrison and Paul Dourish, "Re-Place-ing Space: The Roles of Space and Place in Collaborative Systems," *Proceedings of the ACM conference, Computer Supported Cooperative Work CSCW'96*, Boston, MA (New York: ACM Press, 1996), 70.
77 According to de Certeau, place is unlike space. While place is stable, space is produced, like the spoken word, and caught in the ambiguity of its actualization, just as the city street, defined by the imposed order of its geographical dimensions, is transformed into space by its walkers. See Michel de Certeau, *The Practice of Everyday Life*, 117.
78 Paul Dourish, "Re-Space-ing Place: 'Place' and 'Space' Ten Years On," *Proceedings of the 2006 20th Anniversary Conference on Computer Supported Cooperative Work CSCW '06* (New York: ACM Press, 2006), 304.
79 Locative media are mobile communication systems (supported by Wi-Fi, cellphones, GPS technologies, PDAs, tablet PCs, earphones, etc.), functionally bound to

certain physical locations, able to trigger social activities and urban experiences, by relocating the user from the virtual world of the computer screen to the digitally-mediated physical environment of the city. They can engage playful, situated, and participatory activities in the city, which are often determined not by the designer or media artist, but by users. Locative media technologies can enhance spontaneous social connectivity in the public urban space, creating communities and social relations within both the digital and physical place of individual action. See, for instance, Dimitris Charitos, "Τα Μέσα Επικοινωνίας δι' Εντο☐ισμού και οι Ε☐ιδράσεις τους ως ☐ρος την Κοινωνική Αλληλόδραση στο Περιβάλλον της σημερινής Πόλης," Ζητήματα Ε☐ικοινωνίας 5 (2007): 46–61.

80 The "dérive," literally meaning "drifting," is a central situationist practice. City drifters wander through the city determining transitions of psychological ambiences of attraction or repulsion beyond the control of any central principle or prevalent economic power. The dérive is not a random activity, a simple stroll or a journey, because the drifters will have to let go of their customary work and leisure activities, and be driven by psychological urban encounters that encourage or discourage them to enter certain urban areas. It is a playful-constructive behavior, involving the awareness of psychogeographical effects. See Guy Debord, "Theory of the Dérive," in *Theory of the Dérive and other Situationist Writings on the City*, eds. Libero Andreotti and Xavier Costa (Barcelona: Museu d'Art Contemporani de Barcelona, 1996), 22.

81 The programs of locative media happenings, like the *Conflux Festival* (http://www.confluxfestival.org) and *Come out and Play* (http://www.comeoutandplay.org), emphasize phrases such as "contemporary psychogeography" and "playful-creative behavior." This seems to bring forth situationist strategies for the transformation of the city, although the emphasis now is on digitally-assisted urban games. See Conor McGarrigle, "The Construction of Locative Situations: Locative Media and the Situationist International, Recuperation or Redux?" *UC Irvine: Digital Arts and Culture 2009*, accessed April 5, 2014, http://escholarship.org/uc/item/90m1k8tb.

82 *Tactical Sound Garden* (http://www.tacticalsoundgarden.net) enables users to personalize and customize the soundscape of the city by "planting" a kind of "sound" garden in different locations for other users to download. *You are not Here* (http://www.youarenothere.org) employed two-layered maps of hostile cities – Tel Aviv and Gaza or New York and Baghdad. Users could choose to conceptually visit the other city through their routes in real locations of the first city. Each point of interest on the map had a telephone number that the participants could call, when they were in the actual location, in order to access a voice guide about the corresponding location in the other city. *Walking Tools* (http://www.walkingtools.net) provided a series of open-source software tools that allowed the conversion of mobile phones into open-ended user customizable locative media.

83 For a discussion on this difference from a cognitive psychology perspective see Barbara Tversky, "Structures of Mental Spaces," *Environment and Behavior* 35, 1 (2003): 66–80.

84 Paul Dourish and Genevieve Bell, *Divining a Digital Future: Mess and Mythology in Ubiquitous Computing* (Cambridge, MA/London: MIT Press, 2011), 163.

85 Ibid., 180.

86 Ibid., 183.

87 Thomas Berker, Maren Hartmann, Yves Punie, and Katie Ward, "Introduction," in *Domestication of Media and Technology* (Maidenhead: Open University Press, 2006), 2.

88 See Roger Silverstone, Eric Hirch, and David Morley, "Information and Communication Technologies and the Moral Economy of the Household," in *Consuming Technologies-Media and Information in Domestic Spaces*, eds. Roger Silverstone and Eric Hirsch (London: Routledge, 1992), 15–31. Also see Thomas Berker, Maren Hartmann, Yves Punie and Katie Ward, *Domestication of Media and Technology*.

89 Silverstone, Hirch, and Morley, "Information and Communication Technologies," 21.
90 Knut H. Sørensen, "Domestication: The Enactment of Technology," in *Domestication of Media and Technology*, 46. For instance, the fluidity and variety of family arrangements affects the role played by media in the household. The meaning and significance of household media in different family structures, such as single-parent families, is modified by the complexity of their life, time and financial limitations, and other social and psychological factors. See Anna Maria Russo Lemor, "Making a 'home'. The domestication of information and communication technologies in single parents' households," in *Domestication of Media and Technology*, 165–184.
91 Katie Ward, "The bald guy just ate an orange. Domestication, work and home," in *Domestication of Media and Technology*, 148.
92 Ibid., 149.
93 Ibid., 154.
94 Rapoport, "The Home Under Surveillance," 330–331.
95 Verbeek, *What things do,* 163–168.

BIBLIOGRAPHY

Addington, Michelle, and Daniel Schodek. *Smart Materials and Technologies: For the Architecture and Design Professions.* Oxford: Architectural Press, 2005.

Ahola, Jari. "Ambient Intelligence." *Special Theme: Ambient Intelligence – ERCIM News* 47, (October 2001): 8. http://www.ercim.org/publication/Ercim_News/enw47/intro.html.

Akrich, Madeleine. "The De-scription of Technological Objects." In *Shaping Technology/Building Society: Studies in Sociotechnical Change*, edited by Wiebe Bijker and John Law, 205–224. Cambridge, MA: MIT Press, 1992.

Amsterdamska, Olga. "Surely You're Joking, Mr Latour!" *Science, Technology, Human Values* 15, 4 (1990): 495–504.

Andreotti, Libero, and Xavier Costa eds. *Theory of the Dérive and other Situationist Writings on the City*. Barcelona: Museu d'Art Contemporani de Barcelona, 1996.

Appleby, John. "Planned Obsolescene: Flying into the Future with Stelarc." In *The Cyborg Experiments: The Extensions of the Body in the Media Age*, edited by Joanna Zylinska, 101–113. London: Continuum, 2002.

Ashby, Ross W. "Homeostasis." In *Cybernetics: Transactions of the Ninth Conference*, 73–108. New York: Josiah Macy Foundation, 1952.

Augugliaro, Federico, Sergei Lupashin, Michael Hamer, Cason Male, Markus Hehn, Mark W. Mueller, Jan Willmann, Fabio Gramazio, Matthias Kohler, and Raffaello D'Andrea. "The Flight Assembled Architecture Installation: Cooperative construction with flying machines." *IEEE Control Systems* 34, 4 (2014): 46–64. doi:10.1109/MCS.2014.2320359.

Aztiria, Asier, Alberto Izaguirre, Rosa Basagoiti, Juan Carlos Augusto, and Diane J. Cook. "Discovering Frequent Sets of Actions in Intelligent Environments." In *Intelligent Environments 2009: Proceedings of the 5th International Conference on Intelligent Environments*, edited by Victor Callaghan, Achilleas Kameas, Angélica Reyes, Dolors Royo, and Michael Weber, 153–160. Amsterdam: IOS Press, 2009.

Ball, Mathew, and Vic Callaghan. "Explorations of Autonomy: An Investigation of Adjustable Autonomy in Intelligent Environments." In *2012 8th International Conference on Intelligent Environments (IE)*, Guanajuato, Mexico, 26–29 June 2012, 114–121. IEEE publications, 2012. doi:10.1109/IE.2012.62.

Ball, Mathew, Victor Callaghan, Michael Gardner, and Dirk Trossen. "Exploring Adjustable Autonomy and Addressing User Concerns in Intelligent Environments." In *Intelligent Environments 2009: Proceedings of the 5th International Conference on Intelligent Environments*, edited by Victor Callaghan, Achilleas Kameas, Angélica Reyes, Dolors Royo, and Michael Weber, 429–436. Amsterdam: IOS Press, 2009.

Banham, Reyner. "A Home is not a House." In *Rethinking Technology: A Reader in Architectural Theory*, edited by William W. Braham and Jonathan A. Hale, 159–166. London/New York: Routledge, 2007.

Banham, Reyner. "1960-Stocktaking." In *A critic writes: essays by Reyner Banham*, edited by Mary Banham, Sutherland Lyall, Cedric Price, and Paul Barker, 49–63. Berkeley/Los Angeles CA: University of California Press, 1996.

Banham, Reyner. "A Home is not a House." *Art in America* 2 (1965): 109–118.

Banham, Reyner. "1960 1: Stocktaking-Tradition and Technology." *Architectural Review* 127, 756 (February 1960): 93–100.

Banham, Reyner. *Theory and Design in the first Machine Age*. London: Architectural Press, 1960.

Barker-Plummer, David. "Turing Machines." In *The Stanford Encyclopedia of Philosophy*. Accessed January 10, 2015. http://plato.stanford.edu/archives/sum2013/entries/turing-machine.

Barthes, Roland. "The Death of the Author." In *Image – Music – Text*. Translated by Stephen Heath, 142–148. London: Fontana, 1977.

Bateson, Gregory. *Steps to an Ecology of Mind*. New York: Ballantine Books, 1972.

Beesley, Philip. "Introduction." In *Kinetic Architectures and Geotextile Installations*, edited by Philip Beesley, 19–28. New York: Riverside Architectural Press, 2010.

Beesley, Philip. "Introduction: Liminal Responsive Architecture." In *Hylozoic Ground: Liminal Responsive Architecture*, edited by Philip Beesley, 2–33. New York: Riverside Architectural Press, 2010.

Beesley, Philip, Sachiko Hirosue, and Jim Ruxton. "Toward Responsive Architectures." In *Responsive Architectures: Subtle Technologies*, edited by Philip Beesley, Sachiko Kirosue, Jim Ruxton, Marion Trankle and Camille Turner, 3–11. New York: Riverside Architectural Press, 2006.

Beigl, Michael. "Ubiquitous Computing: Computation Embedded in the World." In *Disappearing Architecture: From Real to Virtual to Quantum*, edited by Georg Flachbart and Peter Weibel, 53–60. Basel: Birkhäuser, 2005.

Bell, Daniel. *The Coming of the Post-Industrial Society*. New York: Basic Books, 2008 [1973].

Berger, Peter, and Thomas Luckmann. *The Social Construction of Reality: A Treatise in the Sociology of Knowledge*. Garden City, NY: Anchor Books, 1966.

Berker, Thomas, Maren Hartmann, Yves Punie and Katie Ward, eds. *Domestication of Media and Technology*. Maidenhead: Open University Press, 2006.

Berryman, Sylvia. "The Imitation of Life in Ancient Greek Philosophy." In *Genesis Redoux: Essays in the History and Philosophy of Artificial Life*, edited by Jessica Riskin, 35–45. Chicago, IL/London: The University of Chicago Press, 2007.

Bier, Henriette, and Terry Knight. *Footprint: Delft School of Design Journal* (*Digitally Driven Architecture*) 6 (2010).

Bijdendijk, Frank. "Solids." In *Time-Based Architecture*, edited by Bernard Leupen, René Heijne and Jasper van Zwol, 42–52. Rotterdam: 010 Publishers, 2005.

Biloria, Nimish. "Introduction: Real Time Interactive Environments. A Multidisciplinary Approach towards Developing Real-Time Performative Spaces." In *Hyperbody. First Decade of Interactive Architecture*, edited by Kas Oosterhuis, Henriette Bier, Nimish Biloria, Chris Kievid, Owen Slootweg, and Xin Xia, 368–381. Heijningen: Jap Sam Books, 2012.

Biloria, Nimish. "Introduction: Hyperbody. Engineering an Innovative Architectural Future." In H*yperbody. First Decade of Interactive Architecture*, edited by Kas Oosterhuis, Henriette Bier, Nimish Biloria, Chris Kievid, Owen Slootweg, and Xin Xia, 176–177. Heijningen: Jap Sam Books, 2012.

Bingham, Geoffrey P., Richard C. Schmidt, and Lawrence D. Rosenblum. "Dynamics and the Orientation of Kinematic Forms in Visual Event Recognition." *Journal of Experimental Psychology: Human Perception and Performance* 21, 6 (1995): 1473–1493.

Bloor, David. "Anti-Latour." *Studies in History and Philosophy of Science* 30, 1 (1999): 81–112. doi:10.1016/S0039-3681(98)00038-7.

Boden, Margaret A., ed. *The Philosophy of Artificial Intelligence*. Oxford/New York: Oxford University Press, 1990.

Bonabeau, Eric, and Guy Théraulaz. "Swarm Smarts." *Scientific American* 282, March 2000, 72–79.

Bonabeau, Eric, Marco Dorigo, and Guy Théraulaz. *Swarm Intelligence: From Natural to Artificial Systems*. New York: Oxford University Press, 1999.

Boyer, Christine M. "The Body in the City: A Discourse on Cyberscience." In *The Body in Architecture*, edited by Deborah Hauptmann. Rotterdam: 010 Publishers, 2006.

Braham, William W., and Paul Emmons. "Upright or Flexible? Exercising Posture in Modern Architecture." In *Body and Building: Essays on the Changing Relation of Body and Architecture*, edited by George Dodds and Robert Tavernor, 290–303. Cambridge, MA/London: MIT Press, 2002.

Braham, William W., and Jonathan A. Hale, eds. *Rethinking Technology: A Reader in Architectural Theory*. London/New York: Routledge, 2007.

Brand, Stewart. *How Buildings Learn: What happens after they're built*. New York: Penguin, 1995.

Breazeal, Cynthia. "Towards Sociable Robots." *Robotics and Autonomous Systems* 42, 3–4 (2003): 167–175.

Breazeal, Cynthia L. "Sociable Machines: Expressive Social Exchange between Robots and People." D.Sc. diss., Massachusetts Institute of Technology, 2000.

Brey, Philip. "Freedom and Privacy in Ambient Intelligence." *Ethics and Information Technology* 7, 3 (2006): 157–166.

Brodey, Warren M. "The Design of Intelligent Environments: Soft Architecture." *Landscape* 17, 1 (1967): 8–12.

Brooks, Rodney A. "A Robust Layered Control System for a Mobile Robot." In *Cambrian Intelligence: The early History of the new AI*, edited by Rodney A. Brooks, 3–26. Cambridge, MA: MIT Press, 1999.

Brooks, Rodney A. "Intelligence without Reason." In *Cambrian Intelligence: The early History of the new AI*, edited by Rodney A. Brooks, 133–159. Cambridge, MA: MIT Press, 1999.

Brooks, Rodney A. *Cambrian Intelligence: The early History of the new AI*. Cambridge, MA: MIT Press, 1999.

Brooks, Rodney A., and Anita M. Flynn. "Fast, Cheap and out of Control: A Robot Invasion of the Solar System." *Journal of the British Interplanetary Society* 42 (1989): 478–485.

Bruce, Alison, Illah Nourbakhsh, and Reid Simmons. "The Role of Expressiveness and Attention in Human-Robot Interaction." Paper presented at the IEEE International Conference on Robotics and Automation – ICRA '02, May, 2002.

Bullivant, Lucy. "Intelligent Workspaces: Crossing the Thresholds." *Architectural Design: 4dspace: Interactive Architecture* 75, 1 (2005): 38–45.

Bullivant, Lucy. "Mediating Devices for a Social Statement: Tobi Schneidler Interactive Architect." *Architectural Design: 4dspace-Interactive Architecture* 75, 1 (2005): 72–78.

Bush, Vannevar. "As We May Think." *The Atlantic Monthly*, July 1, 1945. http://www.theatlantic.com/magazine/archive/1945/07/as-we-may-think/303881/?single_page=true.

Cadigan, Pat. *Synners*. New York: Bantam, 1991.

Calef, Scott. "Dualism and Mind." *The Internet Encyclopedia of Philosophy*. Accessed April 5, 2013. http://www.iep.utm.edu/d/dualism.htm.

Callaghan, Victor, Graham Clarke, and Jeannette Chin. "Some Socio-Technical Aspects of Intelligent Buildings and Pervasive Computing Research." *Intelligent Buildings International Journal* 1, 1 (2009): 56–74.

Callaghan, Victor, Graham Clarke, Martin Colley, Hani Hagras, Jeannette Chin, and Faiyaz Doctor. "Inhabited Intelligent Environments." *BT Technology Journal* 22, 3 (2004): 233–247.

Callon, Michel. "Actor–network Theory: The Market Test." In *Actor Network Theory and After*, edited by John Law and John Hassard, 185–186. Oxford: Blackwell, 1999.

Canguilhem, George. "Machine and Organism." Translated by Mark Cohen and Randall Cherry. In *Incorporations*, edited by Jonathan Crary and Sanford Kwinter, 44–69. New York: Zone Books, 1992.

Canguilhem, George. "Machine et Organisme." In *Connaisance de la Vie*. Paris: Hachette, 1952.

Caporael, Linnda R. "Anthropomorphism and Mechano-morphism: Two Faces of the Human Machine." *Computers in Human Behavior* 2, 3 (1986): 215–234.

"Cedric Price's Generator." *Building Design*, February 23, 1979.

Certeau, Michel de. *The Practice of Everyday Life*. Berkeley: University of California Press, 1984.

Chalk, Warren. "Living, 1990: Archigram Group." *Architectural Design* 69, 1–2 (January–February 1999): iv–v.

Chalk, Warren, Peter Cook, Dennis Crompton, David Greene, Ron Herron, and Michael Webb. *Archigram 3: Towards Throwaway Architecture* (1963).

Charitos, Dimitris. "Τα Μέσα Επικοινωνίας δι' Εντο☐ισμού και οι Επιδράσεις τους ως ☐ρος την Κοινωνική Αλληλόδραση στο Περιβάλλον της σημερινής Πόλης." Ζητήματα Επικοινωνίας 5 (2007): 46–61.

Chasin, Alexandra. "Class and its close relations: Identities among Women, Servants, and Machines." In *Posthuman Bodies*, edited by Judith Halberstram and Ira Livingstone, 73–96. Bloomington: Indiana University Press, 1995.

Chin, Jeannette, Vic Callaghan, and Graham Clarke. "A Programming-by-Example Approach to Customizing Digital Homes." In *Proceedings of the 4th IET International Conference on Intelligent Environments*, Seattle WA, 21–22 July 2008, 1–8. London: IET publications, 2008.

Clark, Andy. *Natural-Born Cyborgs: Minds, Technologies, and the Future of Human Intelligence*. Oxford: Oxford University Press, 2003.

Clark, Andy. *Being There: Putting Brain, Body, and World Together Again*. Cambridge, MA: MIT Press, 1997.

Clark, Andy, and David J. Chalmers. "The Extended Mind." *Analysis* 58, 1 (1998): 7–19. doi: 10.1093/analys/58.1.7.

Clark, Herbert H. *Using Language*. Cambridge: Cambridge University Press, 1996.

Clynes, Manfred E., and Nathan S. Kline. "Cyborgs and Space." In *The Cyborg Handbook*, edited by Chris Hables-Gray, 29–33. London/New York: Routledge, 1995.

Collins, Harry M., and Steve Yearley. "Epistemological chicken." In *Science as Practice and Culture*, edited by Andrew Pickering, 301–326. Chicago: University of Chicago Press, 1992.

Collins, Peter. *Changing Ideals in Modern Architecture, 1750–1950*. Montreal: McGill-Queen's University Press, 1998 [1965].
Colquhoun, Alan. *Modern Architecture*. Oxford/New York: Oxford University Press, 2002.
Colquhoun, Alan. "Plateau Beaubourg." In *Essays in Architectural Criticism: Modern Architecture and Historical Change*, edited by Alan Colquhoun, 110–119. Cambridge, MA/London: MIT Press, 1981.
Cook, Diane, Kyle D. Feuz, and Narayanan C. Krishnan. "Transfer Learning for Activity Recognition: A Survey." *Knowledge and Information Systems* 36, 3 (2013): 537–556. doi:10.1007/s10115-013-0665-3.
Cook, Diane, Narayanan Krishnan, and Parisa Rashidi. "Activity Discovery and Activity Recognition: A New Partnership." *IEEE Transactions on Cybernetics* 43, 3 (2013): 820–828. doi:10.1109/TSMCB.2012.2216873.
Cook, Diane J., and Sajal K. Das. "How Smart are our Environments? An updated look at the state of the art." *Pervasive and Mobile Computing* 3, 2 (2007): 53–73. doi:10.1016/j.pmcj.2006.12.001.
Cook, Peter, ed. *Archigram*. New York: Princeton Architectural Press, 1999.
Cook, Peter. "Within the Big Structure." *Archigram 5: Metropolis* (1964).
Cook, Peter, Warren Chalk, Dennis Crompton, Ron Herron, David Greene, and Michael Webb. *Archigram Seven: Beyond Architecture* (1966).
Cook, Peter, Dennis Crompton, David Greene, Ron Herron, and Geoff Taunton. *Archigram Eight: Popular Pak Issue* (1968).
Cook, Peter, Geoff Taunton, Warren Chalk, Dennis Crompton, David Greene, Ron Herron, Michael Webb, and Envirolab. "Room of 1000 Delights," *Archigram Nine: Fruitiest Yet* (1970).
Cowart, Monica. "Embodied Cognition." *The Internet Encyclopedia of Philosophy*. Accessed April 5, 2013. http://www.iep.utm.edu/e/embodcog.htm.
Croft, Catherine. "Movement and Myth: The Schröder House and Transformable Living." *Architectural Design: The Transformable House* 70, 4 (2000): 10–15.
Debord, Guy. "Theory of the Dérive." In *Theory of the Dérive and other Situationist Writings on the City*, edited by Libero Andreotti and Xavier Costa, 22–27. Barcelona: Museu d'Art Contemporani de Barcelona, 1996 [1956].
Deneubourg, Jean Louis, Serge Aron, Simon Goss, and Jacques Marie Pasteels. "The Self-Organizing Exploratory Pattern of the Argentine Ant." *Journal of Insect Behavior* 3, 2 (1990): 159–168. doi:10.1007/bf01417909.
Dery, Mark. *Escape Velocity: Cyberculture at the End of the Century*. New York: Grove Press, 1996.
Dittrich, Winand H., and Stephen Lea. "Visual Perception of Intentional Motion," *Perception* 23, 3 (1994): 253–268.
Doctor, Faiyaz, Hani Hagras, Victor Callaghan, and Antonio Lopez. "An Adaptive Fuzzy Learning Mechanism for Intelligent Agents in Ubiquitous Computing Environments." *Proceedings of the 2004 World Automation Conference*, vol. 16, Sevilla, Spain, June 28–July 1, 2004, 101–106. IEEE publications, 2004.
Don, Abbe, Susan Brennan, Brenda Laurel, and Ben Shneiderman. "Anthropomorphism: From ELIZA to Terminator 2." In *Proceedings of the SIGCHI Conference on Human Factors in Computing System*, 67–70. Monterey, CA: ACM Press, 1992.
Dorigo, Marco, and Mauro Birattari. "Swarm Intelligence." *Scholarpedia* 2, 9 (2007): 1462. doi:10.4249/scholarpedia.1462.
Dorigo, Marco, Elio Tuci, Vito Trianni, Roderich Gross, Shervin Nouyan, Christos Ampatzis, Thomas Halva Labella, Rehan O'Grady, Michael Bonani, Francesco Mondada. "SWARM-BOT: Design and Implementation of Colonies of Self-assembling

Robots." In *Computational Intelligence: Principles and Practice*, edited by Gary Y. Yen, and David B. Fogel, 103–136. New York: IEEE Computational Intelligence Society, 2006.

Dourish, Paul. "Re-Space-ing Place: 'Place' and 'Space' Ten Years On." *Proceedings of the 2006 20th Anniversary Conference on Computer Supported Cooperative Work CSCW '06*, 299–308. New York: ACM Press, 2006. doi:10.1145/1180875.1180921.

Dourish, Paul. *Where the Action is: The Foundations of Embodied Interaction*. Cambridge, MA/London: MIT Press, 2001.

Dourish, Paul, and Genevieve Bell. *Divining a Digital Future: Mess and Mythology in Ubiquitous Computing*. Cambridge, MA/London: MIT Press, 2011.

Dreyfus, Hubert L. *Being-in-the-World: A Commentary on Heidegger's Being and Time, Division I*. Cambridge, MA: MIT Press, 1991.

Dubberly, Hugh, and Paul Pangaro. "On Modelling: What is Conversation and how can we Design for it?" *Interactions – The Waste Manifesto* 16, 4 (July/August 2009): 22–28. doi: 10.1145/1551986.1551991.

Dubberly, Hugh, and Paul Pangaro. "Cybernetics and Service-Craft: Language for Behavior-Focused Design." *Kybernetes* 36, 9/10 (2007): 1301–1317.

Duffy, Brian R. "Anthropomorphism and the Social Robot." *Robotics and Autonomous Systems* 42, 3–4 (2003): 170–190.

Duffy, Brian R. "Anthropomorphism and Robotics." Paper presented at The Society for the Study of Artificial Intelligence and the Simulation of Behaviour – AISB 2002, Imperial College, England, April 3–5, 2002.

Duffy, Brian R., Gina Joue, and John Bourke. "Issues in Assessing Performance of Social Robots." Paper presented at the 2nd WSEAS International Conference RODLICS, Greece, 25–28 September, 2002.

Dufrenoy, Jean. "Systèmes biologiques servant de modèles pour la technologie." *Cahiers des ingénieurs agronomes*, (June–July 1962): 31–32.

Eastman, Charles. "Adaptive-Conditional Architecture." In *Design Participation: Proceedings of the Design Research Society's Conference*, Manchester, September 1971, 51–57. London: Academy Editions, 1972.

Eddy, Timothy J., Gordon Gallup Jr., and Daniel Povinelli. "Attribution of Cognitive States to Animals: Anthropomorphism in Comparative Perspective." *Journal of Social Issues* 49, 1 (1993): 87–101.

Eisenman, Peter. "Post-Functionalism." In *Architecture Theory since 1968*, edited by Michael K. Hays, 236–239. Cambridge, MA: MIT Press, 1998.

Eng, Kynan. "ADA: Buildings as Organisms." In *Gamesetandmatch Conference Proceedings*. Publikatiebureau Bouwkunde, Faculty of Architecture DUT, 2004. http://www.bk.tudelft.nl/fileadmin/Faculteit/BK/Over_de_faculteit/Afdelingen/Hyperbody/Game_Set_and_Match/GameSetandMatch_1/doc/gsm_I.pdf.

Eng, Kynan, Andreas Bäbler, Ulysses Bernardet, Mark Blanchard, Marcio Costa, Tobi Delbrück, Rodney J. Douglas, Klaus Hepp, David Klein, Jonatas Manzolli, Matti Mintz, Fabian Roth, Ueli Rutishauer, Klaus Wassermann, Andrian M. Whatley, Aaron Wittmann, Reto Wyss, and Paul F.M.J. Verschure. "Ada – Intelligent Space: An Artificial Creature for the Swiss Expo.02." In *Proceedings of the 2003 IEEE/RSJ International Conference in Robotics and Automation*, Taipei 14–19 September 2003, 4154–4159. IEEE conference publications, 2003. doi: 10.1109/ROBOT.2003.1242236.

Eng, Kynan, Rodney J. Douglas, and Paul Verschure. "An Interactive Space That Learns to Influence Human Behavior." *IEEE Transactions on Systems, Man, and Cybernetics – Part A: Systems and Humans* 35, 1 (2005): 66–77.

Eng, Kynan, Matti Mintz, and Paul Verschure. "Collective Human Behavior in Interactive Spaces." In *Proceedings of the 2005 IEEE/RSJ International Conference in Robotics and*

Automation, Barcelona, 18–22 April 2005, 2057–2062. IEEE conference publications, 2005. doi:10.1109/ROBOT.2005.1570416.

Fallan, Kjetil. "De-scribing Design: Appropriating Script Analysis to Design History." *Design Issues* 24, 4 (2008): 61–75.

Fernández-Galiano, Luis. "Organisms and Mechanisms, Metaphors of Architecture." In *Rethinking Technology: A Reader in Architectural Theory*, edited by William Braham and Jonathan Hale, 256–275. London/New York: Routledge, 2007.

Forty, Adrian. *Words and Buildings: A Vocabulary of Modern Architecture*. London: Thames & Hudson, 2004.

Fox, Michael A. "Beyond Kinetic." In *Gamesetandmatch Conference Proceedings*. Publikatiebureau Bouwkunde, Faculty of Architecture DUT, 2004. http://www.bk.tudelft.nl/fileadmin/Faculteit/BK/Over_de_faculteit/Afdelingen/Hyperbody/Game_Set_and_Match/GameSetandMatch_1/doc/gsm_I.pdf.

Fox, Michael A., and Miles Kemp. *Interactive Architecture*. New York: Princeton Architectural Press, 2009.

Fox, Michael A., and Bryant P. Yeh. "Intelligent Kinetic Systems in Architecture." In *Managing Interactions in Smart Environments*, edited by Paddy Nixon, Gerard Lacey and Simon Dobson, 91–103. London: Springer, 2000.

Frazer, John. *An Evolutionary Architecture*. London: Architectural Association, 1995.

Friedman, Yona. "The Flatwriter: choice by computer," *Progressive Architecture* 3, (March 1971): 98–101.

Friedrich, Christian. "Immediate Architecture." In *Hyperbody. First Decade of Interactive Architecture*, edited by Kas Oosterhuis, Henriette Bier, Nimish Biloria, Chris Kievid, Owen Slootweg, and Xin Xia, 225–260. Heijningen: Jap Sam Books, 2012.

Gage, Stephen. "Intelligent Interactive Architecture." *Architectural Design* 68, 11–12 (1998): 80–85.

Garfinkel, Harold. *Studies in Ethnomethodology*. New York: Prentice Hall, 1967.

Gaur, Viksit, and Brian Scassellati. "A Learning System for the Perception of Animacy." Paper presented at the 6th International Conference on Development and Learning, Bloomington, Indiana, 2006.

Gere, Charlie. *Digital Culture*. London: Reaktion Books, 2008 [2002].

Gibson, James J. *The Ecological Approach to Visual Perception*. Hillsdale, NJ: Lawrence Erlbaum Associates, 1986 [1979].

Giedion, Siegfried. "Industrialization as a Fundamental Event." In *Rethinking Technology: A Reader in Architectural Theory*, edited by William Braham and Jonathan Hale, 75–105. London/New York: Routledge, 2007.

Glanville, Ranulph. "An Intelligent Architecture," *Convergence: The International Journal of Research into New Media and Technologies* 7, 2 (2001): 12–24.

Glynn, Ruairi. "Performative Ecologies (2008–10)." *Ruairi Glynn*. Accessed February 2, 2013. http://www.ruairiglynn.co.uk/portfolio/performative-ecologies.

Glynn, Ruairi. "Performative Ecologies." *Ruairi Glynn*. Video, 3:26. Accessed February 2, 2013. http://www.ruairiglynn.co.uk/portfolio/performative-ecologies.

Glynn, Ruairi. "Conversational Environments Revisited." Paper presented at the 19th European Meeting on Cybernetics and Systems Research, Vienna, 25–28 March, 2008.

Gomi, Takashi. "Aspects of non-Cartesian Robotics." *Artificial Life and Robotics* 1, 2 (1997): 95–103.

Goulthorpe, Mark. "Aegis Hyposurface: Autoplastic to Alloplastic." *Architectural Design: Hypersurface Architecture II* 69, 9–10 (1999): 60–65.

Grosz, Elizabeth A. *Time Travels: Feminism, Nature, Power*. Crows Nest: Allen & Unwin, 2005.

Hables-Gray, Chris, ed. *The Cyborg Handbook*. London/New York: Routledge, 1995.
Hables-Gray, Chris, Steven Mentor, and Heidi Figueroa-Sarriera. "Cyborgology: Constructing the Knowledge of Cybernetic Organisms." In *The Cyborg Handbook*, edited by Chris Hables-Gray, 1–14. London/New York: Routledge, 1995.
Habraken, John N. *Supports: An Alternative to Mass Housing*. Translated by B. Valkenburg. London: Architectural Press, 1972.
Hagras, Hani, Victor Callaghan, Martin Colley, Graham Clarke, Anthony Pounds-Cornish, and Hakan Duman. "Creating an Ambient-Intelligence Environment using Embedded Agents." *IEEE Intelligent Systems* 19, 6 (2004): 12–20. doi:10.1109/MIS.2004.61.
Haque, Usman. "Distinguishing Concepts: Lexicons of Interactive Art and Architecture." *Architectural Design: 4dsocial-Interactive Design Environments* 77, 4 (2007): 24–31.
Haque, Usman. "The Architectural relevance of Gordon Pask." *Architectural Design: 4dsocial-Interactive Design Environments* 77, 4 (2007): 54–61.
Haque Design & Research. "Reconfigurable House." Accessed October 12, 2013. http://www.haque.co.uk/reconfigurablehouse.php.
Haraway, Donna J. *Simians, Cyborgs and Women: The Reinvention of Nature*. New York: Routledge, 1991.
Haraway, Donna J. "A manifesto for Cyborgs: Science, Technology and Socialist-Feminism in the 1980s." *Socialist Review* 80, (1985): 65–108.
Harrison, Steve, and Paul Dourish. "Re-Place-ing Space: The Roles of Space and Place in Collaborative Systems." *Proceedings of the ACM conference, Computer Supported Cooperative Work CSCW'96*, Boston, MA, 67–76. New York: ACM Press, 1996. doi:10.1145/240080.240193.
Harvey, David. *Explanation in Geography*. London: Edward Arnold, 1969.
Haugeland, John, ed. *Mind Design: Philosophy, Psychology, Artificial Intelligence*. Cambridge, MA/London: MIT Press, 1981.
Hayles, Katherine N. *How we became Posthuman: Virtual Bodies in Cybernetics, Literature, and Informatics*. Chicago, IL: The University of Chicago Press, 1999.
Hayles, Katherine N. "Life Cycle of Cyborgs." In *The Cyborg Handbook*, edited by Chris Hables-Gray, 321–335. London/New York: Routledge, 1995.
Hays, Michael K., ed. *Architecture Theory since 1968*. Cambridge, MA: MIT Press, 1998.
Heidegger, Martin. *Being and Time*. New York: Harper and Row, 1962 [1927].
Heider, Fritz and Mary-Ann Simmel. "An Experimental Study of Apparent Behavior." *American Journal of Psychology* 57, 2 (1944): 243–259.
Hermann, Haken. "Self-organization." *Scholarpedia* 3, 8 (2008): 1401. doi:10.4249/scholarpedia.1401.
Herron, Ron. "Holographic scene setter." *Archigram Nine: Fruitiest Yet* (1970).
Hertzberger, Herman. *Lessons for Students in Architecture*. Rotterdam: 010 Publishers, 1991.
Heylighen, Francis, and Cliff Joslyn. "Cybernetics and Second Order Cybernetics." In *Encyclopedia of Physical Science & Technology*, vol. 4, edited by Robert Allen Meyers, 155–170. New York: Academic Press, 2001.
Heylighen, Francis, Margeret Heath, and Frank Van Overwalle. "The Emergence of Distributed Cognition: A Conceptual Framework." *Principia Cybernetica Web*. Accessed April 2, 2013. http://pespmc1.vub.ac.be/papers/distr.cognitionframework.pdf.
Hill, Jonathan. *Actions of Architecture: Architects and Creative Users*. London: Routledge, 2003.
Hill, Jonathan, ed. *Occupying Architecture*. London: Routledge, 1998.
Hoberman, Chuck. "Unfolding Architecture." *Architectural Design Profile: Folding in Architecture*, 2004, 72–75.

Hollan, James D., Edwin Hutchins, and David Kirsh. "Distributed Cognition: Toward a new Foundation for Human–computer Interaction Research." *ACM Transactions on Computer-Human Interaction* 7, 2 (June 2000): 174–196. doi:10.1145/353485.353487.

Hosale, Mark David, and Chris Kievid. "InteractiveWall. A Prototype for an E-Motive Architectural Component." In *Hyperbody. First Decade of Interactive Architecture*, edited by Kas Oosterhuis, Henriette Bier, Nimish Biloria, Chris Kievid, Owen Slootweg, and Xin Xia, 484–496. Heijningen: Jap Sam Books, 2012.

Hughes, Jonathan. "The Indeterminate Building." In *Non-Plan: Essays on Freedom Participation and Change in Modern Architecture and Urbanism*, edited by Jonathan Hughes and Simon Sadler, 90–103. Oxford: Architectural Press, 2000.

Hughes, Jonathan, and Simon Sadler, eds. *Non-Plan: Essays on Freedom Participation and Change in Modern Architecture and Urbanism.* Oxford: Architectural Press, 2000.

Hunt, Gillian. "Architecture in the Cybernetic Age." *Architectural Design* 68, 11–12 (1998): 53–55.

Hutchins, Edwin. *Cognition in the Wild*. Cambridge, MA: MIT Press, 1995.

Ihde, Don. *Postphenomenology and Technoscience: The Peking University Lectures*. Albany, NY: State University of New York Press, 2009.

Ihde, Don. *Bodies in Technology*. Minneapolis: University of Minnesota Press, 2002.

Ikonen, Veikko. "Sweet Intelligent Home – User Evaluations of MIMOSA Housing Usage Scenarios." In *The 2nd IET International Conference on Intelligent Environments (IE06)*, vol. 2, Athens, 5–6 July 2006, 13–22. London: IET publications, 2006.

Intille, Stephen S., Kent Larson, Jennifer S. Beaudin, Emmanuel Munguia Tapia, Pallavi Kaushik, Jason Nawyn, and Thomas J. McLeish. "The PlaceLab: a Live-In Laboratory for Pervasive Computing Research (Video)." *Proceedings of Pervasive 2005 Video Program*, Munich, Germany, May 8–13, 2005. Accessed January 27, 2009, http://web.media.mit.edu/~intille/papers-files/PervasiveIntilleETAL05.pdf.

Interactive Institute Smart Studio. "RemoteHome." Accessed May 22, 2014. http://smart.interactiveinstitute.se/smart/projects/remotehome/index_en.html.

Ishii, Hiroshi. "The Tangible User Interface and its Evolution." *Communications of the ACM (Special Issue: Organic user interfaces)* 51, 6 (2008): 32–36.

Ishii, Hiroshi, and Brygg Ullmer. "Tangible Bits: Towards Seamless Interfaces between People, Bits and Atoms." In *Proceedings of the ACM CHI 97 Human Factors in Computing Systems Conference*, edited by Steven Pemberton, 234–241. Atlanta, GA: ACM Press, 1997.

Isozaki, Arata. "Erasing Architecture into the System." In *Re: CP*, edited by Cedric Price and Hans-Ulrich Obrist, 25–51. Basel: Birkhäuser, 2003.

Jaskiewicz, Thomasz. "(In)formed Complexity. Approaching Interactive Architecture as a Complex Adaptive System." In *Hyperbody. First Decade of Interactive Architecture*, edited by Kas Oosterhuis, Henriette Bier, Nimish Biloria, Chris Kievid, Owen Slootweg, and Xin Xia, 184–185. Heijningen: Jap Sam Books, 2012.

Johnston, John. *The Allure of Machinic Life: Cybernetics, Artificial Life, and the New AI.* Cambridge, MA/London: MIT Press, 2008.

Jones, Stephen. "Intelligent Environments: Organisms or Objects?" *Convergence: The International Journal of Research into New Media and Technologies* 7, 2 (2001): 25–33.

Jormakka, Kari. *Flying Dutchmen: Motion in Architecture*. Basel: Birkhäuser, 2002.

Jormakka, Kari, and Oliver Schurer. "The End of Architecture as we know it." In *The 2nd IET International Conference on Intelligent Environments (IE06)*, vol. 2, Athens, Greece, 5–6 July 2006, 105–110. London: IET publications, 2006.

Kalafati, Eleni, and Dimitris Papalexopoulos. *Τάκης Χ. Ζενέτος: Ψηφιακά Οράματα και Αρχιτεκτονική*. Athens: Libro, 2006.

Kameas, Achilles, Irene Mavrommati, and Panos Markopoulos. "Computing in Tangible: Using Artifacts as Components of Ambient Intelligence Environments." In *Ambient Intelligence: The Evolution of Technology, Communication and Cognition Towards the Future of Human–computer Interaction*, edited by Giuseppe Riva, Francesco Vatalaro, Fabrizio Davide, and Mariano Alcaniz, 121–142. Amsterdam/Oxford/Fairfax: IOS Press, 2005.

Kapp, Ernst. *Grundlinien einer Philosophie der Technik*. Braunschweig: Westermann, 1877.

Kaptelinin, Victor, and Bonnie A. Nardi. *Acting with Technology: Activity Theory and Interaction Design*. Cambridge, MA: MIT Press, 2006.

Keller, Evelyn Fox. "Marrying the Premodern to the Postmodern: Computers and Organisms after WWII." In *Prefiguring Cyberculture: an Intellectual History*, edited by Darren Tofts, Annemarie Jonson and Alessio Cavallaro, 52–65. Cambridge, MA/Sydney: MIT Press/Power Publications, 2002.

Kelly, Kevin. *Out of Control: The Rise of Neo-biological Civilization*. Reading, MA: Addison-Wesley, 1994.

Kievid, Chris, and Kas Oosterhuis. "Muscle NSA: A Basis for a True Paradigm Shift in Architecture." In *Hyperbody. First Decade of Interactive Architecture*, edited by Kas Oosterhuis, Henriette Bier, Nimish Biloria, Chris Kievid, Owen Slootweg, and Xin Xia, 445–452. Heijningen: Jap Sam Books, 2012.

Koolhaas, Rem. "The Smart Landscape." *ArtForum*, April 2015. Accessed April 4, 2015. https://artforum.com/inprint/id=50735.

Koolhaas, Rem, Stefano Boeri, Sanford Kwinter, Nadia Tazi, and Hans Ulrich Obrist (eds). *Mutations*. Barcelona: Actar, 2000.

Krementsov, Nikolai, and Daniel Todes. "On Metaphors, Animals, and Us," *Journal of Social Issues* 47, 3 (1991): 67–81.

Kronenburg, Robert. *Portable Architecture*. Oxford: The Architectural Press, 1996.

Kronenburg, Robert. *Flexible: Architecture that Responds to Change*. London: Laurence King Publishing, 2007.

Krueger, Ted. "Eliminate the Interface." In *Gamesetandmatch Conference Proceedings*. Publikatiebureau Bouwkunde, Faculty of Architecture DUT, 2004. http://www.bk.tudelft.nl/fileadmin/Faculteit/BK/Over_de_faculteit /Afdelingen/Hyperbody/Game_Set_and_Match/GameSetandMatch_1/doc/gsm_I.pdf.

Krueger, Ted. "Editorial." *Convergence: The International Journal of Research into New Media and Technologies* 7, 2 (2001): unpaginated.

Krueger, Ted. "Autonomous Architecture." *Digital Creativity* 9, 1 (1998): 43–47.

Krueger, Ted. "Like a second skin, living machines." *Architectural Design* 66, 9–10 (1996): 29–32.

Kuhn, Thomas S. *The Structure of Scientific Revolutions*. Chicago, IL: University of Chicago Press, 1962.

Kuzmanovic, Maja. "Formalising Operational Adaptive Methodologies or Growing Stories within Stories." *AHDS Guides to Good Practice, A Guide to Good Practice in Collaborative Working Methods and New Media Tools Creation*. Accessed June 12, 2014. http://www.ahds.ac.uk/creating/guides/new-media-tools/kuzmanovic.htm.

Lakoff, George, and Mark Johnson. *Philosophy in the Flesh: The Embodied Mind and its Challenge to Western Thought*. New York: Basic Books, 1999.

Landau, Royston. "A Philosophy of Enabling: The Work of Cedric Price." *AA Files* 8 (January 1985): 3–7. http://www.jstor.org/stable/29543432

Langevin, Jared. "Reyner Banham: In Search of an Imageable, Invisible Architecture." *Architectural Theory Review* 16, 1 (2011): 2–21.

Langton, Christopher G., ed. *Artificial Life: The Proceedings of an Interdisciplinary Workshop on the Synthesis and Simulation of Living Systems Held September, 1987 in Los Alamos, New Mexico*, vol. VI. Los Alamos, September 1987. Boston, MA: Addison-Wesley Publishing, 1989.

Langton, Christopher G. "Artificial Life." In *Artificial Life: The Proceedings of an Interdisciplinary Workshop on the Synthesis and Simulation of Living Systems Held September, 1987 in Los Alamos, New Mexico*, vol. VI, edited by Christopher G. Langton, Los Alamos, September 1987, 1–47. Boston, MA: Addison-Wesley Publishing, 1989.

Larson, Kent, and Richard Topping. "PlaceLab: A House_n + TIAX Initiative." House_n. Accessed May 24, 2013. http://architecture.mit.edu/house_n/documents/PlaceLab.pdf.

Latour, Bruno. *Reassembling the Social: an Introduction to Actor–network-Theory*. Oxford: Oxford University Press, 2005.

Latour, Bruno. "Mixing Humans and Nonhumans Together: The Sociology of a Door-Closer." *Social Problems* 35, 3 (1998): 298–310.

Latour, Bruno. *We Have Never Been Modern*. Translated by Catherine Porter. Cambridge, MA: Harvard University Press, 1993.

Latour, Bruno. "Where Are the Missing Masses? The Sociology of a Few Mundane Artifacts." In *Shaping Technology/Building Society: Studies in Sociotechnical Change*, edited by Wiebe Bijker and John Law, 151–180. Cambridge, MA: MIT Press, 1992.

Latour, Bruno. *Science in Action*. Cambridge, MA: Harvard University Press, 1987.

Law, John. "Networks, Relations, Cyborgs: on the Social Study of Technology." *Centre for Science Studies, Lancaster University*. 2000. Accessed March 15, 2013. http://www.comp.lancs.ac.uk/sociology/papers/Law-Networks-Relations-Cyborgs.pdf.

Law, John. "Notes on the Theory of the Actor–network: Ordering, Strategy and Heterogeneity." *Systems Practice* 5, 4 (1992): 379–393.

Lefebvre, Henri. *The Production of Space*. Translated by Donald Nicholson-Smith. Oxford: Blackwell, 1991.

Leontiev, Aleksei N. *Activity, Consciousness, and Personality*. Translated by Marie J. Hall. Hillsdale: Prentice-Hall, 1978.

Leontiev, Aleksei N. "Activity and Consciousness." In *Philosophy in the USSR, Problems of Dialectical Materialism*, 180–202. Moscow: Progress Publishers, 1977.

Leroi-Gourhan, André. *Evolution et techniques II: Milieu et techniques*. Paris: Albin Michel, 1945.

Leslie, Thomas. "Just What is it That Makes Capsule Homes so Different, so Appealing? Domesticity and the Technological Sublime, 1945 to 1975." *Space and Culture* 9, 2 (May 2006): 180–194.

Leupen, Bernard. *Frame and Generic Space*. Rotterdam: 010 Publishers, 2006.

Leupen, Bernard. "Towards Time-Based Architecture." In *Time-Based Architecture*, edited by Bernard Leupen, René Heijne and Jasper van Zwol, 12–20. Rotterdam: 010 Publishers, 2005.

Leupen, Bernard, René Heijne, and Jasper van Zwol, eds. *Time-Based Architecture*. Rotterdam: 010 Publishers, 2005.

Lévy, Pierre. *Cyberculture*. Translated by Robert Bononno. Minneapolis: University of Minnesota Press, 2001.

Lévy, Pierre. *Qu'est-ce que le Virtuel?* Paris: La Decouverte, 1998.

Lévy, Pierre. *Collective Intelligence: Mankind's Emerging World in Cyberspace*. Translated by Robert Bononno. New York: Plenum, 1997.

Lister, Martin, Jon Dovey, Seth Giddings, Iain Grant, and Kieran Kelly. *New Media: A Critical Introduction*. London/New York: Routledge, 2003.

Llewelyn-Davies, Richard. "Endless Architecture." *Architectural Association Journal*, (July 1951): 106–112.

Lobsinger, Mary Lou. "Cedric Price: An Architecture of Performance." *Daidalos* 74 (2000): 22–29.

Logan, Beth, Jennifer Healey, Matthai Philipose, Emmanuel Munguia Tapia, and Stephen Intille. "A Long-Term Evaluation of Sensing Modalities for Activity Recognition." *Proceedings of the International Conference on Ubiquitous Computing*, LNCS 4717, 483–500. Berlin/Heidelberg: Springer-Verlag, 2007.

Massey, Doreen. "A global sense of place." *Marxism Today* 38, (1991): 24–29.

Mathews, Stanley. *From Agit-Prop to Free Space: The Architecture of Cedric Price*. London: Black Dog Publishing, 2007.

Mathews, Stanley. "The Fun Palace as Virtual Architecture: Cedric Price and the Practices of Indeterminacy." *Journal of Architectural Education* 59, 3 (2006): 39–48.

Mathews, Stanley. "Cedric Price: From 'Brain Drain' to the 'Knowledge Economy'." *Architectural Design: Manmade Modular Megastructures* 76, 1 (2005): 90–95.

Maturana, Humberto R., and Francisco J. Varela. *Autopoiesis and Cognition: The Realization of the Living*. Dordrecht/London: Reidel, 1980.

Maudlin, Daniel, and Marcel Vellinga, eds. *Consuming Architecture: On the Occupation, Appropriation and Interpretation of buildings*. London: Routledge, 2014.

Mavrommati, Irene, and Achilles Kameas. "Enabling End Users to compose Ubiquitous Computing applications – The e-Gadgets Experience." *DAISy Dynamic Ambient Intelligent Systems Unit*. Accessed September 8, 2012. http://daisy.cti.gr/webzine/Issues/Issue%202/Articles/Enabling%20end%20users%20to%20compose%20Ubiquitous%20Computing%20applications%20-%20the%20e-Gadgets%20experience/index.html.

Mazlish, Bruce. "The Man-Machine and Artificial Intelligence." *Stanford Electronic Humanities Review* 4, 2 (1995). http://web.stanford.edu/group/SHR/4-2/text/mazlish.html.

Mazlish, Bruce. *The Fourth Discontinuity: The Co-Evolution of Humans and Machines*. New Haven/London: Yale University Press, 1993.

McCarthy, John. "The Little Thoughts of Thinking Machines." *Psychology Today* 17, 12 (1983): 46–49.

McCarthy, John. "Ascribing Mental Qualities to Machines." In *Philosophical Perspectives in Artificial Intelligence*, edited by Martin Ringle, 1–20. Brighton, Sussex: Harvester Press, 1979.

McCorduck, Pamela. *Machines who Think*. San Francisco, CA: W.H. Freeman, 1979.

McCullough, Malcolm. *Digital Ground: Architecture, Pervasive Computing and Environmental Knowing*. Cambridge, MA/London: MIT Press, 2005.

McCulloch, Warren, S., and Walter H. Pitts, "A Logical Calculus of the Ideas Immanent in Nervous Activity." In *Embodiments of Mind*, edited by Warren S. McCulloch, 19–39. Cambridge, MA: MIT Press, 1965. Originally published in *Bulletin of Mathematical Biophysics* 5, (1943): 115–133.

McFarland, David. *Animal Behaviour: Psychobiology, Ethology, and Evolution*. New York: Longman, 1998.

McGarrigle, Conor. "The Construction of Locative Situations: Locative Media and the Situationist International, Recuperation or Redux?" UC Irvine: Digital Arts and Culture 2009. Accessed April 5, 2014. http://escholarship.org/uc/item/90m1k8tb.

McLeod, Mary. "Architecture and Politics in the Reagan Era: From Postmodernism to Deconstructivism." In *Architecture Theory since 1968*, edited by Michael K. Hays, 680–702. Cambridge, MA: MIT Press, 1998.

McLuhan, Marshall. *Understanding Media: the Extensions of Man* (Critical Edition). Corte Madera, CA: Gingko Press, 2003 [1964].
Minsky, Marvin L. *The Society of Mind*. New York: Simon & Schuster, 1985.
Minsky, Marvin L. *Semantic Information Processing*. Cambridge, MA: MIT Press, 1968.
Minsky, Marvin L., and Seymour Papert. *Perceptrons: An Introduction to Computational Geometry*. Cambridge, MA: MIT Press, 1969.
Miranda Carranza, Pablo, and Paul Coates. "Swarm modelling – The use of Swarm Intelligence to generate architectural form." Paper presented at the Generative Art 3rd International Conference, Milano, 2000.
Mitchell, William J. "After the Revolution: Instruments of Displacement." In *Disappearing Architecture: From Real to Virtual to Quantum*, edited by Georg Flachbart and Peter Weibel, 18–23. Basel: Birkhäuser, 2005.
Mitchell, William J. *Me++: The Cyborg Self and the Networked City*. Cambridge, MA/London: MIT Press, 2003.
Monekosso, Dorothy, Paolo Remagnino, and Kuno Yoshinori, eds. *Intelligent Environments: Methods, Algorithms and Applications*. London: Springer, 2008.
Morris, Richard. *Artificial Worlds: Computers, Complexity, and the Riddle of Life*. London/New York: Plenum Trade, 1999.
Müller, Martin. "Assemblages and Actor–networks: Rethinking Socio-material Power, Politics and Space." *Geography Compass* 9, 1 (2015): 27–41, doi:10.1111/gec3.12192.
Mumford, Lewis. "Technical Syncretism and Toward an Organic Ideology." In *Rethinking Technology: A Reader in Architectural Theory*, edited by William Braham and Jonathan Hale, 53–57. London/New York: Routledge, 2007.
"Muscle NSA." *TUDelft*. Accessed August 2, 2014. http://www.bk.tudelft.nl/index.php?id=16070&L=1.
"Muscle Reconfigured." *TUDelft*. Accessed August 2, 2014. http://www.bk.tudelft.nl/index.php?id=16058&L=1.
"Muscle Room." *TUDelft*. Accessed August 2, 2014. http://www.bk.tudelft.nl/en/about-faculty/departments/architectural-engineering-and-technology/organisation/hyperbody/research/applied-research-projects/muscle-room.
"Muscle Tower II: An interactive and Kinetic Tower." *TUDelft*. Accessed July 31, 2014. http://www.bk.tudelft.nl/index.php?id=16060&L=1.
Negroponte, Nicholas. *Soft Architecture Machines*. Cambridge, MA/London: MIT Press, 1975.
Negroponte, Nicholas. "Toward a Theory of Architecture Machines." *Journal of Architectural Education* (1947–1974) 23, 2 (March 1969): 9–12.
Newel, Allen, and Herbert Simon. "Computer Science as Empirical Inquiry: Symbols and Search." *The Tenth Turing Lecture, Communications of the Association for Computing Machinery* 19, 3 (1976): 113–126.
Nieuwenhuys, Constant. "Another City for Another Life." In *Theory of the Dérive and other Situationist Writings on the City*, edited by Libero Andreotti and Xavier Costa, 92–95. Barcelona: Museu d'Art Contemporani de Barcelona, 1996.
Nieuwenhuys, Constant. "New Babylon." In *Theory of the Dérive and other Situationist Writings on the City*, edited by Libero Andreotti and Xavier Costa, 154–169. Barcelona: Museu d'Art Contemporani de Barcelona, 1996.
Nieuwenhuys, Constant. "The Great Game to Come." In *Theory of the Dérive and other Situationist Writings on the City*, edited by Libero Andreotti and Xavier Costa, 62–63. Barcelona: Museu d'Art Contemporani de Barcelona, 1996.
Norman, Donald A. *Living with complexity*. Cambridge, MA/London: MIT Press, 2011.

Norman, Donald A. "THE WAY I SEE IT: Signifiers, not Affordances." *Interactions – Designing Games: Why and How* 15, 6 (November/December 2008): 18–19. doi: 10.1145/1409040.1409044.
Norman, Donald A. *The Design of Future Things*. New York: Basic Books, 2007.
Norman, Donald A. *Things that Make us Smart: Defending Human Attributes in the Age of the Machine*. Boston MA: Addison-Wesley Publishing, 1993.
Norman, Donald A. *The Design of Everyday Things*. New York: Doubleday/Currency, 1990.
Oberquelle, Horst, Ingbert Kupka, and Susanne Maass. "A view of Human–machine Communication and Cooperation." *International Journal of Man-Machine Studies* 19, 4 (1983): 309–333.
Oosterhuis, Kas. "E-motive Architecture." *Inaugural Speech, Delft University of Technology*. Accessed June 24, 2014. http://www.bk.tudelft.nl/en/about-faculty/departments/architectural-engineering-and-technology/organisation/hyperbody/research/theory/e-motive-architecture/2001-e-motive-architecture.
Oosterhuis, Kas. "E-motive House." *ONL [Oosterhuis_Lénárd]*. 2006. Accessed June 24, 2014. http://www.oosterhuis.nl/quickstart/index.php?id=348 [site discontinued].
Oosterhuis, Kas. "The Octogon Interview." *ONL [Oosterhuis_Lénárd]*. 2006. Accessed March 1, 2014. http://www.oosterhuis.nl/quickstart/index.php?id=453.
Oosterhuis, Kas. "A New Kind of Building." In *Disappearing Architecture: From Real to Virtual to Quantum*, edited by Georg Flachbart and Peter Weibel, 90–115. Basel: Birkhäuser, 2005.
Oosterhuis, Kas. *Hyperbodies: Towards an E-motive Architecture*. Basel: Birkhäuser, 2003.
Oosterhuis, Kas. *Architecture Goes Wild*. Rotterdam: 010 Publishers, 2002.
Oosterhuis, Kas. *Programmable Architecture*. Milan: L'Arca Edizioni, 2002.
Oosterhuis, Kas, and Xin Xia, eds. *iA#1 – Interactive Architecture*. Heijningen: Jap Sam Books, 2007.
Oosterhuis, Kas, Henriette Bier, Nimish Biloria, Chris Kievid, Owen Slootweg, and Xin Xia, eds. *Hyperbody. First Decade of Interactive Architecture*. Heijningen: Jap Sam Books, 2012.
Oungrinis, Konstantinos. *Μεταβαλλόμενη Αρχιτεκτονική: Κίνηση, Προσαρμογή, Ευελιξία*. Athens: ION publications, 2012.
Oungrinis, Konstantinos. *Δομική Μορφολογία και Κινητικές Κατασκευές στους Μεταβαλλόμενους Χώρους*. PhD diss., Aristotelian University of Thessaloniki, 2009.
Oungrinis, Konstantinos. *Transformations: Paradigms for Designing Transformable Spaces*. Cambridge, MA: Harvard University Graduate School of Design, 2006.
Oungrinis, Konstantinos-Alketas, and Marianthi Liapi. "Spatial Elements Imbued with Cognition: A possible step toward the 'Architecture Machine'." *International Journal of Architectural Computing* 12, 4 (2014): 419–438.
Palumbo, Maria Luisa. *New Wombs: Electronic Bodies and Architectural Disorders*. Basel: Birkhäuser, 2000.
Papalexopoulos, Dimitris. *Ψηφιακός Τοπικισμός*. Athens: Libro, 2008.
Pask, Gordon. "The Limits of Togetherness." In *Information Processing 80: Congress Proceedings*, edited by Simon H. Lavington, 999–1012. Amsterdam: North-Holland Publishing Company, 1980.
Pask, Gordon. *Conversation Theory, Applications in Education and Epistemology*. New York: Elsevier, 1976.
Pask, Gordon. *Conversation, Cognition and Learning*. New York: Elsevier, 1975.
Pask, Gordon. *The Cybernetics of Human Learning and Performance*. London: Hutchinson Educational, 1975.
Pask, Gordon. "A comment, a case history and a plan." In *Cybernetics, Art and Ideas*, edited by Jasia Reichardt, 89–99. London: Studio Vista, 1971.

Pask, Gordon. "The Architectural Relevance of Cybernetics." *Architectural Design* 9 (1969): 494–496.
Pask, Gordon. "The Colloquy of Mobiles." In *Cybernetic Serendipity: the computer and the arts*, edited by Jasia Reichardt, 34–35. London/New York: Studio International, 1968.
Pask, Gordon. "A plan for an aesthetically potent social environment." *Archigram Eight: Popular Pak Issue* (1968).
Pfaffenberger, Bryan. "Actant." *STS Wiki*. Accessed April 22, 2013. http://www.stswiki.org/index.php?title=Actant.
Pfeifer, Rolf, and Bongard, Josh C. *How the Body Shapes the Way We Think: A New View of Intelligence*. Cambridge, MA: MIT Press, 2006.
Pickering, Andrew. *The Cybernetic Brain: Sketches of another Future*. Chicago, IL/London: University of Chicago Press, 2009.
Pickering, Andrew. *The Mangle of Practice: Time, Agency, and Science*. Chicago: University of Chicago Press, 1995.
Pickering, Andrew. "The mangle of practice: agency and emergence in the sociology of science." *American Journal of Sociology* 99, 3 (1993): 559–589. doi:10.1086/230316.
Picon, Antoine. *Digital Culture in Architecture: An Introduction for the Design Professions*. Basel: Birkhauser, 2010.
Picon, Antoine. "Architecture, Science and Technology." In *The Architecture of Science*, edited by Peter Galison and Emily Thompson, 309–335. Cambridge, MA/London: MIT Press, 1999.
Picon, Antoine. *La Ville Territoire des Cyborgs*. Besançon: Editions de l'Imprimeur, 1998.
Picon, Antoine. "Towards a History of Technological Thought." In *Technological Change. Methods and themes in the history of technology*, edited by Robert Fox, 37–49. London: Harwood Academic Publishers, 1996.
Poster, Mark. "High-Tech Frankenstein, or Heidegger Meets Stelarc." In *The Cyborg Experiments: The Extensions of the Body in the Media Age*, edited by Joanna Zylinska, 15–32. London: Continuum, 2002.
Preston, John. *Kuhn's The Structure of Scientific Revolutions: A Reader's Guide*. London/New York: Continuum, 2008.
Price, Cedric. *Cedric Price. Works II*. London: Architectural Association, 1984.
Rabeneck, Andrew. "Cybermation: A Useful Dream." *Architectural Design* (September 1969): 497–500.
Ramachandran, Vilayanur S., and William Hirstein. "The Science of Art: A Neurological Theory of Aesthetic Experience." *Journal of Consciousness Studies* 6, 6–7 (1999): 15–51.
Rapoport, Michele. "The Home Under Surveillance: A Tripartite Assemblage." *Surveillance & Society* 10, 3–4 (2012): 320–333. http://library.queensu.ca/ojs/index.php/surveillance-and-society/article/view/tripartite/tripartite.
Reconfigurable House "Hacking Low Tech architecture." Accessed October 12, 2013. http://house.propositions.org.uk.
Risan, Lars Christian. "Artificial Life: A Technoscience Leaving Modernity? An Anthropology of Subjects and Objects." *AnthroBase*. Accessed December 12, 2013. http://www.anthrobase.com/Txt/R/Risan_L_05.htm.
Riskin, Jessica, ed. *Genesis Redux: Essays in the History and Philosophy of Artificial Life*. Chicago, IL/London: The University of Chicago Press, 2007.
Riskin, Jessica. "Eighteenth-Century Wetware." *Representations* 83, (2003): 97–125.
Riskin, Jessica. "The Defecating Duck, or, the Ambiguous Origins of Artificial Life." *Critical Inquiry* 29, 4 (2003): 598–633.
Rivas, Jesus and Gordon Burghardt. "Crotalomorphism: A Metaphor to Understand Anthropomorphism by Omission." In *The Cognitive Animal: Empirical and Theoretical*

Perspectives on Animal Cognition, edited by Marc Bekoff, Colin Allen, and Gordon Burghardt, 9–17. Cambridge, MA: MIT Press, 2002.

Rivera-Illingworth, Fernando, Victor Callaghan, and Hani Hangras. "A Neural Network Agent Based Approach to Activity Detection in AmI Environments." *IEE Seminar on Intelligent Building Environments* 2, 11059 (2005): 92–99.

Robbin, Tony. *Engineering a New Architecture*. New Haven, CT: Yale University Press, 1996.

Rogers, Richard. Postscript to *Supersheds: The Architecture of Long-Span Large-Volume Buildings*, edited by Chris Wilkinson. Oxford: Butterworth-Heinemann, 1991.

Rogers, Yvonne. "Distributed Cognition and Communication." In *The Encyclopaedia of Language and Linguistics*, edited by Keith Brown, 181–202. Oxford: Elsevier, 2006.

Roosegaarde, Daan. "Liquid Constructions." In *GameSetandMatch II: On Computer Games, Advanced Geometries and Digital Technologies*, edited by Kas Oosterhuis and Lukas Feireiss, 164–170. Rotterdam: Episode Publishers, 2006.

Rumelhart, David, and James McClelland. *Parallel Distributed Processing*, Vol. 1–2. Cambridge, MA: MIT Press, 1986.

Russel, Stuart J., and Peter Norvig. *Artificial Intelligence: a Modern Approach*. New Jersey: Pearson, 2003 [1995].

Russo Lemor, Anna Maria. "Making a 'home'. The domestication of Information and Communication Technologies in single parents' households." In *Domestication of Media and Technology*, edited by Thomas Berker, Maren Hartmann, Yves Punie, and Katie Ward, 165–184. Maidenhead: Open University Press, 2006.

Sack, Warren. "Artificial Human Nature." *Design Issues* 13, 2 (1997): 55–64.

Sacks, Harvey. *Lectures on Conversation*, vol. 1–2. Oxford: Blackwell, 1995.

Sacks, Harvey, Emanuel A. Schegloff, and Gail Jefferson. "A Simplest Systematics for the Organization of Turn-Taking for Conversation." In *Studies in the Organization of Conversational Interaction*, edited by Jim Schenkein, 7–55. New York: Academic Press, 1978.

Sadler, Simon. *Archigram: Architecture without Architecture*. Cambridge, MA/London: MIT Press, 2005.

Sadler, Simon. "Open Ends: The Social Visions of 1960s non-Planning." In *Non-Plan: Essays on Freedom Participation and Change in Modern Architecture and Urbanism*, edited by Jonathan Hughes, and Simon Sadler, 138–154. Oxford: Architectural Press, 2000.

Sadler, Simon. *The Situationist City*. Cambridge, MA: MIT Press, 1998.

Saffiotti, Alessandro, Mathias Broxvall, Marco Gritti, Kevin LeBlanc, Robert Lundh, Jayedur Rashid, BeomSu Seo, and Young-Jo Cho. "The PEIS-Ecology Project: Vision and Results." In *Proceedings of the IEEE/RSJ International Conference on Intelligent Robots and Systems, IROS 2008*, Nice, France, September 2008, 2329–2335. IEEE publications, 2008.

Saggio, Antonino. *The IT Revolution in Architecture: Thoughts on a Paradigm Shift*. Translated by Stephen Jackson. New York: lulu.com, 2013 [2007].

Saggio, Antonino. "Architecture and Information Technology: A Conceptual Framework." In *The 2nd IET International Conference on Intelligent Environments (IE06)*, vol. 1, Athens, 5–6 July 2006, 9–12. London: IET publications, 2006.

Saggio, Antonino. "How." In *Behind the Scenes: Avant-Garde Techniques in Contemporary Design*, edited by Francesco De Luca, and Marco Nardini, 5–7. Basel: Birkhäuser, 2002.

Saggio, Antonino. "New Crisis, New Researchers: Interactivity vs Transparency, Subjectivity vs Objectivity, Desires vs Needs." Prima Facoltà di Architettura – Roma La Sapienza. Accessed May 19, 2003. http://www.arc1.uniroma1.it/saggio/conferenze/lo/4d.htm.

Sandhana, Lakshmi. "Smart Buildings Make Smooth Moves." *Wired*, August 31, 2006. Accessed August 23, 2009. http://archive.wired.com/science/discoveries/news/2006/08/71680.

Schmid, Christian. "Henri Lefebvre's Theory of the Production of Space: Towards a three-dimensional dialectic.' In *Space, Difference, Everyday Life: Reading Henri Lefebvre*, edited by Kanishka Goonewardena, Stefan Kipfer, Richard Milgrom, and Christian Schmid, 27–45. London/New York: Routledge, 2008.

Schneider, Tatjana, and Jeremy Till. *Flexible Housing*. Oxford: Architectural Press, 2007.

Scholl, Brian J., and Patrice D. Tremoulet. "Perceptual Causality and Animacy." *Trends in Cognitive Sciences* 4, 8 (2000): 299–309.

Scimemi, Maddalena. "The Unwritten History of the Other Modernism: Architecture in Britain in the Fifties and Sixties." *Daidalos* 74 (2000): 15–20.

Searle, John. "Minds, Brains, and Programs." *Behavioural and Brain Sciences* 3, 3 (1980): 417–424.

Serres, Michel. *Le Parasite*. Paris: Grasset, 1980.

Shapiro, Stuart C., ed. *Encyclopedia of Artificial Intelligence*. New York: Wiley, 1987.

Silverstone, Roger, Eric Hirch, and David Morley. "Information and Communication Technologies and the Moral Economy of the Household." In *Consuming Technologies-Media and Information in Domestic Spaces*, edited by Roger Silverstone and Eric Hirsch, 15–31. London: Routledge, 1992.

Singla, Geetika, and Diane J. Cook, "Interleaved Activity Recognition for Smart Home residents." In *Intelligent Environments 2009: Proceedings of the 5th International Conference on Intelligent Environments*, edited by Victor Callaghan, Achilleas Kameas, Angélica Reyes, Dolors Royo, and Michael Weber, 145–152. Amsterdam: IOS Press, 2009.

Slootweg, Owen. "Muscle Tower II." In *Hyperbody. First Decade of Interactive Architecture*, edited by Kas Oosterhuis, Henriette Bier, Nimish Biloria, Chris Kievid, Owen Slootweg, and Xin Xia, 404–408. Heijningen: Jap Sam Books, 2012.

Sofoulis, Zoë. "Cyberquake: Haraway's Manifesto." In *Prefiguring Cyberculture: an Intellectual History*, edited by Darren Tofts, Annemarie Jonson, and Alessio Cavallaro, 84–103. Cambridge, MA/Sydney: MIT Press/Power Publications, 2002.

Sørensen, Knut H. "Domestication: The Enactment of Technology." In *Domestication of Media and Technology*, edited by Thomas Berker, Maren Hartmann, Yves Punie, and Katie Ward, 40–61. Maidenhead: Open University Press, 2006.

Spigel, Lynn. "Designing the Smart House: Posthuman Domesticity and Conspicuous Production." *European Journal of Cultural Studies* 8, 4 (2005): 403–426.

Spiller, Neil. *Visionary Architecture: Blueprints of the Modern Imagination*. London: Thames & Hudson, 2006.

Stamatis, Panagiotis, Ioannis Zaharakis, and Achilles Kameas. "Exploiting Ambient Information into Reactive Agent Architectures." In *The 2nd IET International Conference on Intelligent Environments* (IE06), vol. 1, Athens, 5–6 July 2006, 71–78. London: IET publications, 2006.

Stanek, Łukasz. *Henri Lefebvre on Space: Architecture, Urban Research, and the Production of Theory*. Minneapolis/London: University of Minnesota Press, 2011.

Steiner, Hadas A. "Off the Map." In *Non-Plan: Essays on Freedom Participation and Change in Modern Architecture and Urbanism*, edited by Jonathan Hughes and Simon Sadler, 126–137. Oxford: Architectural Press, 2000.

Steiner, Hadas A. *Beyond Archigram: The Structure of Circulation*. London/New York: Routledge, 2009.

Stelarc. "Obsolete Bodies." *Stelarc*. Accessed March 2, 2014. http://stelarc.org/?catID=20317.

Stelarc. "Third Hand." *Stelarc*. Accessed March 2, 2014. http://stelarc.org/?catID=20265.
Stelarc. "Fractal Flesh." *Stelarc*. Accessed March 2, 2014. http://stelarc.org/?catID=20317.
Stewart, Judith Ann. "Perception of Animacy." PhD diss., University of Pennsylvania, 1982.
Stiegler, Bernard. *La Technique and le temps, vol. 2, La Désorientation*. Paris: Galilée, 1996.
Streitz, Norbert A. "Designing Interaction for Smart Environments: Ambient Intelligence and the Disappearing Computer." In *The 2nd IET International Conference on Intelligent Environments (IE06)*, vol. 1, Athens, 5–6 July 2006, 3–8. London: IET publications, 2006.
Suchman, Lucy A. *Human–machine Reconfigurations: Plans and Situated Actions*. Cambridge: Cambridge University Press, 2007.
Suchman, Lucy A. *Plans and Situated Actions: The Problem of Human Machine Communication*. Cambridge: Cambridge University Press, 1987.
Sudjic, Deyan. "Birth of the Intelligent Building." *Design*, January, 1981.
Sullivan, C.C. "Robo Buildings: Pursuing the Interactive Envelope." *Architectural Record* 194, 4 (2006), 148–156.
Thackara, John. *In the Bubble: Designing in a Complex World*. Cambridge, MA/London: MIT Press, 2005.
Thacker, Eugene. "Networks, Swarms, Multitudes: part one." *CTheory, 2004*. Accessed February 2, 2013. http://www.ctheory.net/articles.aspx?id=422.
Thacker, Eugene. "Networks, Swarms, Multitudes: part two." *CTheory, 2004*. Accessed February 2, 2013. http://www.ctheory.net/articles.aspx?id=423.
"The Digital House." *Haririandhariri*. Accessed December 2, 2014. http://www.haririandhariri.com.
Thompson, D'Arcy Wentworth. *On Growth and Form*. Cambridge: Cambridge University Press, 1917.
Till, Jeremy. *Architecture Depends*. Cambridge, MA/London: MIT Press, 2009.
Tofts, Darren. "On Mutability." In *Prefiguring Cyberculture: an Intellectual History*, edited by Darren Tofts, Annemarie Jonson, and Alessio Cavallaro, 2–5. Cambridge, MA/Sydney: MIT Press/Power Publications, 2002.
Tremoulet, Patrice D., and Jacob Feldman. "Perception of Animacy from the Motion of a Single Object." *Perception* 29, 8 (2000): 943–951.
Trocchi, Alexander. *Invisible Insurrection of a Million Minds: a Trocchi Reader*. Edinburgh: Polygon, 1991.
Turing, Alan. "Computing Machinery and Intelligence." *Mind* LIX, 236 (October 1950): 433–460. doi:10.1093/mind/LIX.236.433.
Turing, Alan. "On Computable Numbers with Application to the Entscheidungsproblem." *Proceedings of the London Mathematical Society* 42, 2 (1936): 230–265. doi:10.1112/plms/s2-42.1.230.
Turkle, Sherry. *The Second Self: Computers and the Human Spirit*. Cambridge, MA/London: MIT Press, 2005.
Turkle, Sherry. *Life on the Screen: Identity in the age of the Internet*. London: Weidenfeld & Nicolson, 1995.
Tversky, Barbara. "Structures of Mental Spaces." *Environment and Behavior* 35, 1 (2003): 66–80. doi:10.1177/0013916502238865.
Vandenberghe, Frédéric. "Reconstructing Humans: A Humanist Critique of Actant-Network Theory." *Theory, Culture and Society* 19, 5–6 (2002): 55–67.
Varela, Francisco J. *Invitation aux sciences cognitives*. Paris: Éditions du Seuil, 1996.
Varela, Francisco J. *Principles of Biological Autonomy*. New York: Elsevier North Holland, 1979.

Varela, Francisco J., Evan Thompson, and Eleanor Rosch. *The Embodied Mind: Cognitive Science and Human Experience.* Cambridge, MA: MIT Press, 1992.
Verbeek, Peter-Paul. *What things do: Philosophical Reflections on Technology, Agency and Design.* University Park, PA: Pennsylvania State University Press, 2005.
Vidler, Anthony. "Cities of Tomorrow." *Artforum International*, Sept 1, 2012.
von Foerster, Heinz. "On Self-Organizing Systems and Their Environment." In *Observing Systems*. Salinas CA: Intersystems Publications, 1981 [1960].
von Foerster, Heinz. *The Cybernetics of Cybernetics.* Minneapolis: FutureSystems Inc, 1995 [1975].
Vygotsky, Lev. *Mind in Society: The Development of Higher Psychological Processes.* Cambridge, MA: Harvard University Press, 1978.
Walby, Catherine. "The Instruments of Life: Frankenstein and Cyberculture." In *Prefiguring Cyberculture: an Intellectual History*, edited by Darren Tofts, Annemarie Jonson, and Alessio Cavallaro, 28–37. Cambridge, MA/Sydney: MIT Press/Power Publications, 2002.
Ward, Katie. "The bald guy just ate an orange. Domestication, work and home." In *Domestication of Media and Technology*, edited by Thomas Berker, Maren Hartmann, Yves Punie, and Katie Ward, 145–164. Maidenhead: Open University Press, 2006.
Warwick, Kevin. *I, Cyborg.* Champaign, IL: University of Illinois Press, 2004.
Watt, Stuart. "Seeing this as People: Anthropomorphism and Common-Sense Psychology." PhD diss., The Open University, 1998.
Weber, Bruce "Life." 2008. *The Stanford Encyclopedia of Philosophy.* Accessed September 7, 2008. http://plato.stanford.edu/archives/fall2008/entries/life.
Weeks, John. "Indeterminate Architecture." *Transactions of the Bartlett Society* 2, (1962–64): 83–106.
Weinstock, Mike. "Terrain Vague: Interactive Space and the Housescape." *Architectural Design: 4dspace-Interactive Architecture* 75, 1 (2005): 46–50.
Weiser, Mark. "The Computer of the 21st Century." *Scientific American* 265, 3 (1991): 66–75.
Weiser, Mark, Rich Gold, and John Seely Brown. "Origins of Ubiquitous Computing Research at PARC in the Late 1980s." *IBM Systems Journal* 38, 4 (1999): 693–696. doi:10.1147/sj.384.0693.
Wellner, Peter. "Interacting with Paper on the DigitalDesk." *Communications of the ACM* 36, 7 (1993): 87–96.
Wertsch, James, ed. *The Concept of Activity in Soviet Psychology.* Armonk, NY: M.E. Sharpe, 1981.
Whiteley, Nigel. *Reyner Banham: Historian of the Immediate Future.* Cambridge, MA: MIT Press, 2002.
Whittle, Andrea, and André Spicer. "Is actor–network theory critique?" *Organization Studies* 29, 4 (2008): 611–629. doi:10.1177/0170840607082223.
Wiener, Norbert. *The Human Use of Human Beings: Cybernetics and Society.* London: Free Association Books, 1989 [1950].
Wiener, Norbert. *Cybernetics or Control and Communication in the Animal and the Machine.* Cambridge, MA: The Technology Press, 1948.
Wigley, Mark. "Anti-Buildings and Anti-Architects." *Domus* 866, (2004): 15–16.
Wigley, Mark. "Network Fever." *Grey Room* 4, (Summer 2001): 82–122.
Wigley, Mark. "Paper, Scissors, Blur." In *The Activist Drawing: Retracing Situationist Architectures from Constant's New Babylon to Beyond*, edited by Catherine de Zegher and Mark Wigley, 26–56. New York/Cambridge, MA: Drawing Center/ MIT Press, 2001.
Wigley, Mark. *Constant's New Babylon: The Hyper Architecture of Desire.* Rotterdam: Witte de With Center for Contemporary Art/010 Publishers, 1998.

Wilken, Rowan. "Calculated Uncertainty: Computers, Chance Encounters, and 'Community' in the Work of Cedric Price." *Transformations: Accidental Environments* 14, (2007). Accessed August 13, 2014. http://transformationsjournal.org/journal/issue_14/article_04.shtml.

Winner, Langton. "Upon Opening the Black Box and Finding It Empty: Social Constructivism and the Philosophy of Technology." *Science, Technology, & Human Values* 18, 3 (1993): 362–378.

Woolgar, Steve. "Why Not a Sociology of Machines? The Case of Sociology and Artificial Intelligence." *Sociology* 19, 4 (1985): 557–572. doi:10.1177/0038038585019004005.

"World's First Intelligent Building." *RIBA journal.* January, 1982.

Yiannoudes, Socrates. "An Application of Swarm Robotics in Architectural Design." In *Intelligent Environments 2009: Proceedings of the 5th International Conference on Intelligent Environments*, edited by Victor Callaghan, Achilleas Kameas, Angélica Reyes, Dolors Royo, and Michael Weber, 362–370. Amsterdam: IOS Press, 2009.

Yiannoudes, Socrates. "Kinetic Digitally-Driven Architectural Structures as 'Marginal' Objects – A Conceptual Framework." *Footprint – Delft School of Design Journal: Digitally Driven Architecture* 6 (2010): 41–54.

Yiannoudes, Socrates. "The Archigram Vision in the Context of Intelligent Environments and its Current Potential." In *The 7th International Conference on Intelligent Environments (IE)*, Nottingham, 25–28 July 2011, 107–113. London: IEEE publications, 2011.

Yiannoudes, Socrates. "From Machines to Machinic Assemblages: a conceptual distinction between two kinds of Adaptive Computationally-Driven Architecture." In *International Conference Rethinking the Human in Technology-Driven Architecture*, Technical University of Crete, Chania, 30–31 August 2011, 149–162, edited by Maria Voyatzaki and Constantin Spyridonidis. EAAE Transactions on Architectural Education No.55, 2012.

Zaharakis, Ioannis, and Achilles Kameas. "Emergent Phenomena in AmI Spaces." *The European Association of Software Science and Technology Newsletter* 12, (2006): 82–96.

Zaharakis, Ioannis, and Achilles Kameas. "Social Intelligence as the Means for achieving Emergent Interactive Behaviour in Ubiquitous Environments." In *Human–computer Interaction, Part II*, vol. LNCS 4551, edited by Julie Jacko, 1018–1029. Berlin/Heidelberg: Springer-Verlag, 2007.

Zeki, Semir. *Inner Vision: an Exploration of Art and the Brain.* Oxford: Oxford University Press, 2000.

Zenetos, Takis. "Town Planning and Electronics." *Architecture in Greece annual review* 8, (1974): 122–135.

Zenetos, Takis. "City Planning and Electronics." *Architecture in Greece annual review* 3, (1969): 114–125.

Zenetos, Takis. *Urbanisme électronique: structures parallèles / City Planning and Electronics: Parallel Structures / Ηλεκτρονική πολεοδομία: παράλληλες κατασκευές.* Athens: Architecture in Greece Press, 1969.

Zuk, William, and Roger Clark. *Kinetic Architecture.* New York: Van Nostrand Reinhold, 1970.

INDEX

Bold page numbers indicate figures.

A-Life. *see* artificial life (A-Life)
Activity, Consciousness, and Personality (Leontiev) 154–5
activity recognition in intelligent environments 72–9, **77, 78**
activity theory 154–7
actor-network theory (ANT) 151–2, 153–4, 167n112
ADA - The Intelligent Space 165n69; animate and anthropomorphic aspects 95; interactivity 141–4
adaptability as soft flexibility 16
adaptation: architectural perspective 4; computing perspective 4; concept of 4, 5, 9; early references to 4–5; kinetic structures 38–9
Addington, Michelle 70
affordances, theory of 158
agency in interactivity 147–51, **148, 149,** 156–7, 162
Ahola, Jari 70–1
allopoiesis 14
alterity, relations of 159–60
Ambient Agoras research program 187
ambient intelligence: activity recognition 72–9, **77, 78**; biological references 101; context sensing 72–4; control, loss of 80; disappointment in 80; early ideas 3–4; iDorm 73–4; as a layer of a building 189–90, **190**; *PEIS Ecology* 76–7, **77, 78**, 79; *PlaceLab* 74–6;

privacy and security issues 80; social issues 90n91; social problems 79–81; space, place and inhabitation in 188–94, 198
animacy, perception of 94–100, **96**
Animaris Percipiere (Jansen) 97, **97**
Animaris Rhinoceros (Jansen) 97, **98**
anthropomorphism 94–100, **96**
Appleby, John 147–8
Archigram group 19–23, **20, 21, 22,** 43, 101, 186
"Architectural Relevance of Cybernetics, The (Pask) 29
architecture: common interests with computing 4; and homeostasis 11–13, **12**
Architecture of the Well-Tempered Environment, The (Banham) 24
Aristotle 117n54
artificial intelligence: activity recognition 72–9, **77, 78**; and the built environment 59–60; connectionism 58, 59–60; context sensing 72–4; early attempts 55–6, **56, 57,** 57–9; embodied cognition 60–1; functional capabilities 71–2, **72**; General Problem Solver 57–8; *Generator* 55–6, **56, 57**; and human intelligence 58–9; iDorm 73–4; intelligence, meaning of 70–1; intelligent environments 69–71, **70, 72**; machine-organism boundaries

105–6; *PEIS Ecology* 76–7, **77, 78,** 79; *PlaceLab* 74–6; social issues 79–81; subsumption architecture 60–1, **61,** 85n30; swarm intelligence 61–4, **64, 65,** 66, **66, 67, 68**; symbolic 58
artificial life (A-Life) 105, 106, 110, 111, 112, 113, 118n68, 151, 157
Ascott, Roy 29
Ashby, Ross 10
associative indexing 131
asymmetry in interactivity 154–7, 162
augmented artifacts 173–4
automata 103–4, 117n54
autonomous intelligent environments 81–3
autopoiesis 13–15, **15,** 44n13, 105
Autopoiesis and Cognition: The Realization of the Living (Maturana and Varela) 13–14
avant-garde, architectural: adaptability 16; Archigram group 19–23, **20, 21, 22**; biological references 101; Cedric Price 25–9, **27,** 31, **32,** 33; flexibility **15,** 15–17; Fun Palace (Price) 31, **32,** 33; open-ended spaces as aim of 18; Reyner Banham 23–5, **24, 25**; socio-cultural context 18–19; space, theory of 16–17; Takis Zenetos 33–4, **35, 36, 37, 38**; users, conceptualization of 17; Yona Friedman 29–31, **30**

background relations 160
Ball, Matthew 82
Banham, Reyner 1, 23–5, **24, 25,** 47n39, 49n61
Bateson, Gregory 146
Beesley, Philip 106–9, **107, 108, 109,** 118n74
Bell, Genevieve 193
Bengt Sjostrom Starlight Theatre **40, 41**
Biloria, Nimish 92
biology: machine-organism exchanges 102–10, **104, 107, 108, 109**; references to in architecture 100–2, 116n40; and technology, boundary between 103–10, **104, 107, 108, 109**
Blocksworld 13
Borgmann, Albert 160
Boyer, Christine 44n13
Braham, William 102
Brand, Stuart 47n37, 189
Breazeal, Cynthia 104, **104**

Brey, Philip 79
Brodey, Warren 101
Brooks, Rodney 60–1

Callon, Michael 151
Canguilhem, Georges 9–10, 42, 44n3
Caporael, Linnda 94
Carranza, Pablo Miranda 63–4, **64**
Chasin, Alexandra 90n91
Clark, Andy 149
Clark, Roger 38
closed-system kinetics 41–2, 53n114
cognition, embodied 60–1
Colloquy of Mobiles (Pask) 135–7
Colquhoun, Alan 20
component-based architecture 175
computing, common interests with architecture 4
connectionism 58, 59–60
Constant 51n87, 186
context sensing 72–4
contextualization in interactive architecture. *see* interactivity
conversational interaction 132–44, **138, 139**
Cook, Diane 69–70
cultural aspects: biological references in architecture 100–2; machine-organism exchanges 102–10, **104, 107, 108, 109**; nature-culture divide in modernity 110–12
cybernetics: Archigram group 19–23, **20, 21, 22**; autopoiesis 13–15, **15**; basic principles of 10–11; Cedric Price 25–9, **27,** 31, **32, 33**; defined 10; Fun Palace (Price) 28–9, 31, **32,** 33; homeostasis 11; machine-organism boundaries 105; Reyner Banham 23–5, **24, 25**; second-order 13–15, **14,** 29; self-regulation 11; socio-cultural context 18–19; Takis Zenetos 33–4, **35, 36, 37, 38**; Yona Friedman 29–31, **30**. *see also* avant-garde, architectural
"Cyborg Manifesto, A" (Haraway) 145–6
cyborgs 144–51, **148, 149**

4D-Pixel (Roosegaarde) 122–3, **123, 124**
D'Andrea, Raffaello 66, **67, 68**
Das, Sajal 69–70
de Certeau, Michel 17, 192
de Vaucanson, Jacques 103–4
Defecating Duck (De Vaucanson) 103–4

Design of Everyday Things, The (Norman) 158
Design of Intelligent Environments, The (Brodey) 101
desire and need, differentiation between 42–3
distributed cognition 153–4, 170n141
domestication of technology 193–4
Dourish, Paul 157–9, 160–1, 172, 174, 191–2, 193
Duffy, Frank 189

e-gadgets 175–8, 182–6
e-motive architecture 1
e-motive house 8n4, 92–4
Eastman, Charles 4, 11
"Electronic Urbanism" (Zenetos) 33–4, **35, 36, 37**, 38
embodied cognition 60–1
embodied interaction 157–8
Emmons, Paul 102
enaction, theory of 85n21
end-user driven intelligent environments 81–3; augmented artifacts 173–4; e-gadgets 182–6; *Extrovert-Gadgets* 175–8; functionality of 182–; tangible computing and 172–82, **179, 180, 181**
Eng, Kynan 95
engagement with technological devices 160–1
Environment Bubble (Banham) 23, 25
everyday life, theory of 17, 47n33
extended mind theory 168n127

Flatwriter (Friedman) 29–31, **30**
flexibility: and adaptability 16; in archite cture 5; closed-system kinetics 41–2, 53n114; Fun Palace (Price) 26, 28; hard **15,** 15–16; interactive machine, architecture as 18; in modernism **15,** 15–16; as political strategy 17; soft 16; by technical means **15,** 15–16
Flight Assembled Architecture (Gramazio, Kohler and D'Andrea) 66, **67, 68**
Flow 5.0 (Roosegaarde) 122, **125**
"Fluidic Muscle" 52n108
Forty, Adrian 15, 16, 100, 116n40
Fox, Michael 11–12, 38, 39, 121
Fractal Flesh (Stelarc) 147, **148**
Frame and Generic Space (Leupen) 46n28
Frazer, John 56

Friedman, Yona 29–31, **30**
Fun Palace (Price) 25–9, **27,** 31, **32,** 33, 49n64, 49n66, 50n71, 51n86

General Problem Solver 57–8
Generator 55–6, **56, 57**
generic space 46n28
Gibson, James Jerome 158
Glanville, Ranulph 71, 140
Glynn, Ruairi 137, **138, 139,** 139–41
Gramazio, Fabio 66, **67, 68**
Growth and Form (art exhibition) 19

Hamilton, Richard 19
Handswriting (Stelarc) **149**
Haque, Usman 122, 129, 137, 178–82, **179, 180, 181**
Haraway, Donna 105, 106, 145–6, 149, 165n79
Harrison, Steve 192
Hayles, Katherine N. 106, 146, 150, 166n82
Heider, Fritz 95–6
Hertzberger, Herman 16
"Home is Nor a House, A" (Banham) 1, 23–4
homeostasis: and architecture 11–13, **12;** cybernetics 11; machine-organism boundaries 105
Hosale, Mark David 93–4
House_n 74–6
How We Became Posthuman (Hayles) 146
human conversation 132–4
human intelligence and AI 58–9
human-machine interaction. *see* interactivity
Human Use of Human Beings, The (Wiener) 10
Hutchins, Edwin 153, 169n129
Hylozoic Ground (Beesley) 106–9, **107, 108, 109,** 118n74
Hyperbody 1, 62, 92–3, 97–9, 122, 123
Hyposurface 94

iDorm 73–4
Ihde, Don 159–60, 174
In Touch (Tangible Media Group) 187
Independent Group 19
inhabitation, place and space in ambient intelligence 188–94, 198
intelligent environments: activity recognition 72–9, **77, 78;** AI and the built environment 59–60; ambient intelligence 3–4, 72–81,

77, 78; artificial intelligence 55–69, **56, 57, 61, 64, 65, 66, 67, 68**; autonomous 81–3; connectionism 58, 59–60; context sensing 72–4; control, loss of 80; disappointment in 80; early ideas 3–4; end-user driven 81–3; failure of 90n94; functional capabilities 71–2, **72**; iDorm 73–4; intelligence, meaning of 70–1; intelligent environments 69–71, **70, 72**; *PEIS Ecology* 76–7, **77, 78**, 79; *PlaceLab* 74–6; privacy and security issues 80; second-order cybernetics 15; social issues 79–81, 90n91; subsumption architecture 60–1, **61**, 85n30; swarm intelligence 61–4, **64, 65**, 66, **66, 67, 68**; symbolic AI 58
intelligent user interfaces 69
Inter-Action Center (Price) 51n86
InteractiveWall 93–4, 123, 126, 129, **129, 130**, 163n15
interactivity: activity theory 154–7; actor-network theory (ANT) 151–2, 153–4, 167n112; *ADA - The Intelligent Space* 141–4, 165n69; agency in 147–51, **148, 149**, 156–7, 162; alterity, relations of 159–60; in architecture 121–3, 126–30; asymmetry 154–7, 162; autonomous to relational systems 145–51, **148, 149**; background relations 160; conversational interaction 132–44, **138, 139**; cyborgs 144–51, **148, 149**; digital media 131–2; distributed cognition 153–4, 170n141; embodied interaction 157–8; engagement with technological devices 160–1; *InteractiveWall* 123, 126, 129, **129, 130**; interface design 162; mediated relations 159, 160; mediation, philosophy of 159–60; non-verbal 187; *Performative Ecologies* (Glynn) 137, **138, 139**, 139–41; phenomenology 158–9; posthuman 146–7, 166n82; postphenomenological approach 159–61; recognition of intelligence in systems 140; and responsive architecture 122–3, **123, 124, 125**; roots of 130–1; senspsonsive system 126–7; *Spirit/Ghost* **127, 128**; subjectivity, new 145–6; symmetry 153–4; and wider cultural developments 132
interfaces: anthropomorphism and design of 96; intelligent user 69; interactivity and design of 162; open potential environments 197; tangible user interface (TUI) 182
Isozaki, Arata 50n71

Jansen, Theo 97, **97, 98**
Jones, Stephen 102
Jormakka, Kari 63

Kalafati, Eleni 34
Kameas, Achilles 175, 182, 183, 184, 200n35
Kaptelinin, Victor 155–7
Kelly, Kevin 60
Kemp, Miles 121
Kievid, Chris 63, 93–4
kinetic structures: adaptation 38–9; Bengt Sjostrom Starlight Theatre **40, 41**; closed-system kinetics 41–2, 53n114; defined 38; examples **40**, 40–1, **41**; information technology, potential of 42–3; from mechanical to adaptive systems 42–3; mechanisms **39**
Kismet (Breazeal) 104, **104**
Kohler, Matthias 66, **67, 68**
Koolhaas, Rem 94, 189
Krueger, Ted 71
Kuhn, Thomas 118n80

Langton, Christopher 110
Latour, Bruce 110–11, 149, 151, 152
Law, John 151–2
layers of buildings 189–90, **190**
Le Corbusier 15, 45n17, 100
Lefebvre, Henri 16–17, 42–3
Leontiev, Aleksei 154–5
Leupen, Bernard 46n28, 189
Levy, Jody 94
Lévy, Pierre 131
Liapi, Marianthi 99, 126–7
Lister, Martin 110, 147
literature on computing in the built environment 3
Living with Complexity (Norman) 162
locative media 192

machine, man-as, metaphor of 46n21
machine-human interaction. *see* interactivity

machine-organism exchanges 102–10, **104, 107, 108, 109**
Maisons Loucheur 15, 45n18
man-as-machine, metaphor of 46n21
Mangle of Practice, The (Pickering) 164n54
marginal objects 103–10, **104, 107, 108, 109**, 140
maritime navigation 169n129
Mathews, Stanley 28, 31, 49n64, 49n66
Maturana, Humberto 13–14, 105
Mazlish, Bruce 110, 112
McCullough, Malcolm 188–9
McLuhan, Marshall 146
mediated relations 159, 160
mediation, philosophy of 159–60
mental states, changes of 101
Mitchell, William 146–7
Mixing Humans and Nonhumans Together: The Sociology of a Door Closer (Latour) 111
modernism, flexibility in **15**, 15–16
modernity, nature-culture divide in 110–12
multistability of technologies 174
Mumford, Lewis 44n3
Muscle projects 122; *Muscle Body* 41; *Muscle NSA* 40, 62–3; *Muscle Reconfigured* 40–1, 53n109; *Muscle Room* 41; *Muscle Space* 41; *Muscle Tower II* 97–9
Musicolor machine (Pask) 135, 136

Nardi, Bonnie 155–7
nature-culture divide in modernity 110–12
need and desire, differentiation between 42–3
Negroponte, Nicholas 1, 11, 13
neural networks 58, 60
New Babylon (Nieuwenhuys) 51n87, 186
Newell, Allen 57–8
Nieuwenhuys, Constant 51n87, 186
non-verbal interactivity 187
Norman, Donald 134–5, 153, 158, 162

object-subject distinctions 111
Oosterhuis, Kas 1–2, 62, 63, 92–3, 98, 122
open potential environments: agency of users 197–8; alternative capacities of 186–8; e-gadgets 182–6; failure of system, potential 182–3; functionality of end-user driven environments 182; interfaces and middleware 197; living in 186–94, **190**; social contact, mediation of 187–8
organism-machine exchanges 102–10, **104, 107, 108, 109**
Oungrinis, Konstantinos 61, 86n34, 99, 126–7
Oxford Corner House (Price) 51n86

Palumbo, Maria Luisa 59–60, 101–2
Papalexopoulos, Dimitris 34, 187–8, 190
Pask, Gordon 4, 29, 31, 135–7
PEIS Ecology 76–7, **77, 78,** 79
perceptron 58
Performative Ecologies (Glynn) 137, **138, 139,** 139–41, 144
permanence 189
phenomenology 158–9
Pickering, Andrew 31, 105, 136, 137, 164n54
Picon, Antoine 100–1, 109–10
Pillar of Information 29
PiP (Pervasive interactive Programming) 81–2
place, space and inhabitation in ambient intelligence 188–94, 198
PlaceLab 74–6
Plans and Situated Action: The Problem of Human Machine Communication (Suchman) 134
polyvalence 16
Poster, Mark 146, 148
posthuman 146–7, 166n82
postphenomenological approach to interactivity 159–61, 170n161
Price, Cedric 25–9, **27**, 31, **32,** 33, 49n66, 50n71, 51n86, 186–7
privacy issues in intelligent environments 80
Production of Space, The (Lefebvre) 16–17
Project Cyborg (Warwick) 149
Prosthetic Head (Stelarc) 167n107
psychological aspects: animacy, perception of 94–100, **96**; anthropomorphism 94–100, **96**; e-motive house 92–4

Rabeneck, Andrew 4
Rapoport, Michele 80–1, 191
Reconfigurable House (Haque and Somlai-Fischer) 178–82, **179, 180, 181**

RemoteHome project (Schneidler) 187
responsive architecture 1, 122–3, **123, 124, 125**
Risan, Lars 111
Riskin, Jessica 104
robotic systems: anthropomorphism 96; behavior-based 60–1; sociable robots 104, **104**; swarm intelligence in 64
Rogers, Richard 39
Roosegaarde, Daan 122–3, **123, 124, 125**

Sack, Warren 106
Saggio, Antonino 42
Schneider, Tatjana 16
Schneidler, Tobi 187
Schodek, Daniel 70
Schröder Huis 15, 45n16
Schrödinger, Erwin 105
science and architecture, relations between 100–1
scientific management 46n21
Scimemi, Maddalena 50n71
second-order cybernetics 13–15, **14**, 29
security issues in intelligent environments 80
SEEK 13
self-regulation: and architecture 11–13, **12**; cybernetics 11
sensponsive architecture 99, 126–7
Simmel, Mary-Ann 95–6
Simon, Herbert 57–8
situational interaction 132–44, **138, 139**
Snooks, Roland 63
sociable robots 104, **104**, 140
social computing 158–9
social contact, mediation of 187–8
Sofoulis, Zoë 145, 149–50
Soft Architecture Machines (Negroponte) 1
space, place and inhabitation in ambient intelligence 188–94, 198
space, theory of 16–17
Spiegel, Lynn 90n91
Spinal Body Carrier (Zenetos) 34
Spirit/Ghost 99, 127, **127, 128**
Stanek, Lukasz 43
Steiner, Hadas 20–1
Stelarc 147–9, **148, 149**, 166n96, 167n107
'Stocktaking-Tradition and Technology' (Banham) 49n61

Strandbeest (Jansen) 97, **97, 98**
Streitz, Norbert 4
Stuart-Smith, Robert 63
subject-object distinctions 111
subject-object interactions. *see* interactivity
subsumption architecture 60–1, **61**, 85n30
Suchman, Lucy 90n91, 111–12, 132–4, 150–1
surgical implant technology 149, 166n100
swarm intelligence 61–4, **64, 65**, 66, **66, 67, 68**, 190
Swarm-Roof 64, **65**, 66, **66**
symmetry in interactivity 153–4

tactile interactivity 187
tangible computing: augmented artifacts 173–4; defined 172–3; designers, role and presence of 174–5; and end-user driven environments 172–82, **179, 180, 181**; everyday world 158–9; *Extrovert-Gadgets* 175–8; *Reconfigurable House* 178–82, **179, 180, 181**; social contact, mediation of 187–8; tangible user interface (TUI) 182. *see also* open potential environments
tangible user interface (TUI) 182
Taylorism 46n21
TGarden 150, 167n106
Thackara, John 72, 90n94
theoretical framework, lack of 3
Theory and Design in the First Machine Age (Banham) 47n39
Things that make us Smart (Norman) 153
Third Hand (Stelarc) 148, **149**, 166n96
Till, Jeremy 16, 46n21
time and motion studies 46n21
Turing machines 50n83
Turing Test 84n5
Turkle, Sherry 102–3, 104, 116n51

ubiquitous communication 69
ubiquitous computing 69
un-house 1, 23
user interfaces. *see* interfaces
users, conceptualization of 17

Van den Broek, Johannes 15, 45n16
Varela, Francisco 13–14, 85n21, 105
Verbeek, Peter-Paul 159, 160–1, 174
Ville Spatiale (Firedman) 29–30, **30**

Ward, Katie 194
Warwick, Kevin 149, 166n100
Watt, Stuart 94
What Things Do (Verbeek) 159
"Where are the missing masses?" (Latour) 152
Where the Action is: The Foundations of Embodied Interaction (Dourish) 157–9, 160–1

Wiener, Norbert 10, 11
Wood, Robin McKinnon 136
Wright, Frank Lloyd 100

Yeh, Bryant 38, 39

Zaharakis, Ioannis 183, 184, 200n35
Zenetos, Takis 33–4, **35, 36, 37,** 38
Zuk, William 38

Helping you to choose the right eBooks for your Library

Add Routledge titles to your library's digital collection today. Taylor and Francis ebooks contains over 50,000 titles in the Humanities, Social Sciences, Behavioural Sciences, Built Environment and Law.

Choose from a range of subject packages or create your own!

Benefits for you

- Free MARC records
- COUNTER-compliant usage statistics
- Flexible purchase and pricing options
- All titles DRM-free.

Benefits for your user

- Off-site, anytime access via Athens or referring URL
- Print or copy pages or chapters
- Full content search
- Bookmark, highlight and annotate text
- Access to thousands of pages of quality research at the click of a button.

REQUEST YOUR FREE INSTITUTIONAL TRIAL TODAY

Free Trials Available
We offer free trials to qualifying academic, corporate and government customers.

eCollections – Choose from over 30 subject eCollections, including:

Archaeology	Language Learning
Architecture	Law
Asian Studies	Literature
Business & Management	Media & Communication
Classical Studies	Middle East Studies
Construction	Music
Creative & Media Arts	Philosophy
Criminology & Criminal Justice	Planning
Economics	Politics
Education	Psychology & Mental Health
Energy	Religion
Engineering	Security
English Language & Linguistics	Social Work
Environment & Sustainability	Sociology
Geography	Sport
Health Studies	Theatre & Performance
History	Tourism, Hospitality & Events

For more information, pricing enquiries or to order a free trial, please contact your local sales team:
www.tandfebooks.com/page/sales

The home of Routledge books

www.tandfebooks.com